Temporal GIS: Advanced Functions for Field-Based Applications

Springer

Berlin
Heidelberg
New York
Barcelona
Hong Kong
London
Milan
Paris
Tokyo

George Christakos · Patrick Bogaert · Marc L. Serre

Temporal GIS: Advanced Functions for Field-Based Applications

With 73 Figures and 23 Tables

Authors

George Christakos, Ph.D., P.Eng.
Professor & Director
Center for the Advanced Study
of the Environment - CASE
Department of Environmental
Sciences & Engineering
School of Public Health
University of North Carolina
Chapel Hill, NC 27599-7431, USA

Patrick Bogaert, Dr.
Associate Professor
Université Catholique de Louvain
Faculté des Sciences Agronomiques
Place Croix du Sud, 2 Bte 16,
1348 Louvain-la-Neuve, Belgium

Marc L. Serre, Ph.D.
Research Assistant Peofessor
Department of Environmental Sciences
& Engineering, School of Public Health
University of North Carolina
Chapel Hill, NC 27599-7431, USA

ISBN 3-540-41476-2 Springer-Verlag Berlin Heidelberg New York

Library of Congress Cataloging-in-Publication Data applied for

Die Deutsche Bibliothek – CIP Einheitsaufnahme
Christakos, George: Advanced functions of temporal GIS : with 23 tables / George Christakos ; Patrick Bogaert ; Marc L. Serre. -- Berlin ; Heidelberg ; New York ; Barcelona ; Hong Kong ; London ; Milan ; Paris ; Tokyo :Springer, 2002
ISBN 3-540-41476-2

Springer-Verlag Berlin Heidelberg New York
a member of the BertelsmannSpringer Science+Business Media GmbH
http://www.springer.de

Printed in Germany

Cover Design: *design & production*, Heidelberg
Dataconversion: Büro Stasch, Bayreuth (*www.stasch.com*)

SPIN: 10717285 – 5 4 3 2 1 0 – Printed on acid-free paper

Preface

Trust only movement.
Life happens at the level of events not of words.
Trust movement.
A. Adler

As its title suggests, the main goal of this book is the development of *advanced functions* for field-based *Temporal Geographical Information Systems* (*TGIS*). These fields may describe a variety of natural, epidemiological, economical, and social phenomena distributed across space and time. Within such a framework, the book makes an attempt to establish links between, (*a*) the currently conceived *TGIS* techniques, and (*b*) the Bayesian maximum entropy (*BME*) techniques of *Modern Spatiotemporal Geostatistics*. This link could be vital for offering significant improvements in the advanced functions of *TGIS* analysis and modelling, as well as generating useful information in a variety of real-world decision making and planning situations.

To achieve the above goals, the eight Chapters of the book are organized around four main themes:

Concepts, mathematical tools, computer programs, and applications.

In fact, the focus is mainly on the step-by-step implementation of the computational *BME* approach and the extensive use of illustrative examples and real-world applications. Indeed, because of the applied character of the present book, no detailed theoretical explanations or mathematical derivations are included. Instead, the reader is referred to the earlier book by Christakos (*Modern Spatiotemporal Geostatistics*, Oxford Univ. Press, New York, N.Y., 2000) for a comprehensive presentation of these *BME* aspects. With this in mind, the chapter-by-chapter organization of the book is described next.

Chapter 1 reviews the conceptual framework of the *TGIS* considered in this book from a *BME* point of view. Chapter 2 introduces the reader to the fundamental ideas and methods of spatiotemporal analysis, including space/time geometries and random field theory. Chapter 3 discusses issues of physical knowledge synthesis. Chapter 4 presents the reader with the state-of-the-art mathematical tools of spatiotemporal mapping with emphasis on formal *BME* techniques. Comparisons are made with other techniques, like statistical regression, Kriging, neural networks, and Kalman-Bucy filters. Chapter 5 analyzes the interpretive features of *BME*, establishing correspondence relationships between the natural system and the formal mathematics which describe it, measuring and testing the formal structure, or justifying the methodological steps of spatiotemporal mapping. In Chapters 4 and 5 the reader will also find some interesting non-Bayesian and non-entropic extensions of the *BME* approach. The purpose of Chapters 6 and 7 is to make the reader familiar with the *BME* toolbox and the associated library of comprehensive computer programs (*BMElib*) currently used in *Modern Spatiotemporal Geostatistics* applications. This familiarity is achieved by means of analytical examples and numerical experiments, as well as real-world applications. The reader can reconstruct many of the numerical experiments with the help of the *BMElib*, as is indicated by the computer mouse logo (🖱) at the beginning of the relevant examples. We have spent a considerable amount of time on these two chapters and tried to make the material as easy as possible to grasp. Finally, Chapter 8 discusses important uses of the *TGIS* analysis and modelling in the context of scientific hypothesis testing, explanation, and decision making.

The focus on *critical practicality* throughout the book is necessary in order to make the point about the usefulness and relevance of the *BME* techniques in the study of *TGIS* environments. The term "critical" refers to a situation in which practical approaches and software packages are not seen merely as "data massaging" tools, but rather as the means of interpreting and processing knowledge in order to increase one's scientific understanding of the phenomenon of interest, provide accurate predictions, and make useful decisions. We ask the reader to bear with us at some parts of the book where we philosophize about the *TGIS* process. We are convinced that the reward for such an effort is, indeed, substantial. It is hoped that by integrating the current *TGIS* software with the *BME* software presented in this book one could get a whole that is greater than the sum of its parts. The mathematics of the presentation have been kept at a level with which most *TGIS* modellers should feel comfortable with. Some more advanced material, which the reader can omit in a first reading, are identified by the water-clock logo (⌛). Nevertheless, the book contains a number of mathematical equations, for which we make no apology. In sciences, these equations correspond to real physical, biological, ecological, and epidemiological situations and constitute a common language between the various scientific disciplines involved. If *TGIS* specialists really expect to communicate with natural scientists and engineers and make meaningful contributions to the relevant scientific disciplines, they should have considerable familiarity with the basic equations of these disciplines.

The research presented in the book has been partially funded by grants from the National Aeronautics and Space Administration (Grant 60-00RFQ041), the National Institute of Environmental Health Science (Grant P42 ES05948-02), and the U.S. Civilian Research and Development Foundation (Grant RG2-2236). To these financial benefactors we remain grateful. The comments made by Drs. Kyung-mee Choi and Alexander Kolovos were very helpful and appreciated. We are also indebted to Lara Freeburg Kees and Lucinda Thompson for their editorial assistance.

Working on this book has been a very rewarding experience for us. In fact, the book was written while each one of us was working in a different continent: America (George Christakos), Europe (Patrick Bogaert), and Africa (Marc L. Serre). This provides the advanced *TGIS* functions of this book with a kind of an "inter-continental" flavor.

George Christakos Chapel Hill, USA
Patrick Bogaert Louvain-la-Neuve, Belgium
Marc L. Serre Alexandria, Egypt

To the unended quest.
GC, PB, MLS

Contents

Thoughts without content are empty,
intuitions without concepts are blind.
I. Kant

Chapter 1

A *BME* View to the New Realities of *TGIS*

> The totality of our so-called knowledge or beliefs,
> from the most casual matters of geography and history
> to the profoundest laws of atomic physics or even
> of pure mathematics and logic, is a man-made fabric
> which impinges on experience only along the edges
> *W. V. O. Quine*

1.1 Introducing a Temporal Geographical Information System (*TGIS*)

Commonly used atemporal *geographical information systems* (*GIS*) neglect essential dynamics of the natural processes in time and do not take into account important cross-correlations and causal dependencies in the composite space/time domain, thereby significantly limiting the practitioner's understanding of the situation of interest and reducing the predictability capabilities of the system. The goal of this introductory chapter is to present a conceptual overview of *temporal* GIS (*TGIS*) which can describe a variety of phenomena in life support sciences (i.e., natural, social, and health sciences) as well as interactions between these phenomena across space and time. Despite their great potential, there is a general consensus among experts that *TGIS* have not been studied to the extent they deserve (e.g., Langran 1992; Laurini and Thompson 1995; Hamilton and Viscusi 1999; Haggett 2000). Langran, e.g., notices that:

> "While researchers have not neglected spatiotemporal issues, the work that exists suggests only a bare sketch of a *TGIS*", and "fully spatiotemporal attempts either lack a conceptual framework of geographical change or are geared to nongeographic applications altogether".

As a consequence, the *TGIS* definitions occasionally adopted are rather *ad hoc* and limited in scope ignoring salient aspects of real-world spatiotemporal phenomena (see, e.g., Sinton 1978; Bennett 1979; Dangermond 1984; MacEachren and Taylor 1994; Birkin et al. 1996; Mynett 1999; Forgionne et al. 2001). In order to revisit this important issue in a more systematic manner, let us focus our attention on describing some *broad characteristics* of *TGIS*.

1.1.1 Purposefulness, Content, and Context

In general, an important characteristic of a *TGIS* is that it is a *purposeful* system, i.e., it is designed to achieve certain purposes as defined by the user. Furthermore, the phenomena described by *TGIS* may take place within the Earth's limits (e.g., the region within which the human population lives) or outside these geographical limits in Space. The latter is the reason that some authors prefer to refer to spatial, not just geographical systems (Laurini and Thompson 1995). We prefer to use the general term "geographotemporal", which includes phenomena that may occur within the Earth's limits or in Space. In this book, then, we adopt a rather broad definition of *TGIS*:

The combination of scientific modelling and information technology in order to process knowledge about phenomena that occur in a geographotemporal domain and to satisfy particular user needs in the most efficient way.

This definition assigns both a context (i.e., what framework is considered) and a content (i.e., what is included) to the term "*TGIS*". "Information technology" refers to the use of computers for information acquisition, storage, distribution and presentation in order to satisfy specific user needs. While the definition above clearly emphasizes the vital role of computers in the study of natural phenomena and social processes, it also recognizes the importance of the conceptual and logical components of a science-based *TGIS*. The latter includes the use of the appropriate concepts, models, principles, schemes, methods and technology. In short,

Constructing a TGIS *is as much an intellectual as a practical activity.*

In other words, the aim of a *TGIS* is twofold:

i. To have a *content* that satisfies well-established scientific standards of knowledge processing; as well as
ii. To demonstrate its application potential within a specified *context*.

These aspects are at the center of discussions concerning the future of Information Technology (IT).

> Without context, information does not exist, and the context in question must relate not only to the data's environment (where it came from, why it's being communicated, how it's arranged, etc.), but also from the context and intent of the person interpreting it,

says Nathan Shedroff, one of the IT gurus (see, Wurman 2001, p. 28). Furthermore, the above *TGIS* definition is appropriate for the goals of this book for the additional reason that it links *TGIS* with all relevant scientific disciplines which aim at processing various forms of spatiotemporal knowledge (time-georeferenced data, physical laws coordinated in space and time, ecologic patterns, epidemiologic distributions, social processes, etc.). This issue is discussed next.

1.1.2 Synthesis, Organization, and Visualization

TGIS is an interdisciplinary subject that can be considered in various application domains, including environmental monitoring and decision-making, health risk assessment, ecologic hypotheses testing, geocoding, weather forecasting, global warming, mineral exploration, remote sensing, cartography, geodesy, photogrammetry, computer-aided design, transportation planning, economic development, military operations, gender studies, retail planning, and resource management, to mention only a few. What all the domains mentioned above have in common is

The basic need for synthesizing, organizing and visualizing spatiotemporal knowledge sources in a scientifically sound and technologically efficient manner.

Knowledge *synthesis*, *organization* and *visualization* technologies can offer the *TGIS* user with valuable information on a variety of issues (scientific, managerial, administrative, political, etc.). The more effective the synthesis of knowledge from various sources is, the greater the potential impact of *TGIS* to life support sciences will be.[1] Physical knowledge *organization* is as important as creating and acquiring it.

Example 1.1. In environmental science, *TGIS* fulfills the important functions of *connecting* site-specific data to physical models of environmental fate and transport, as well as *visualizing* such a connection. Advanced forms of visualization in space/time will, naturally, include animated maps displaying geographical changes over the time of the processes or objects in question (e.g., geographical maps of radioactive soil contamination continuously updated in time). In the health sciences, the *TGIS*-generated maps of the geographotemporal distribution of a disease (e.g., childhood brain cancer) may show a particular pattern, suggesting the possible role of environmental factors (e.g., water quality). In ecology, maps describing the space/time pattern of abundaces for certain species (e.g., the Eastern Kingbird) offer vital pieces of information regarding the decline of species diversity (e.g., the geographical distribution of abundance may play a role in assessing the risk of extinction for certain species). Finally, the goal of many new disciplines, like bioinformatics, is to find ways to synthesize and process knowledge, not just acquire it. ▪

1.1.3 Action-Oriented

The function of *TGIS* is not restricted to knowledge synthesis; it can also *act* upon the space/time information thus generated on the basis of the appropriate *epistemic* interpretation[2]. In fact, depending on the application environment considered (e.g., the physical situation in which the *TGIS* will be used), the *TGIS* can have several functions, other than traditional information processing. More specifically, the *TGIS*-generated maps can offer spatiotemporal information that is used to

1. Test scientific *hypotheses* and *theories*.
2. *Explain* essential aspects of the phenomenon of interest and *understand* the underlying mechanisms.
3. Improve the effectiveness of *decision* making.

A more detailed discussion of these topics is given in Chap. 8. Naturally, activities (*1*), (*2*), and (*3*) assume a close interaction between the *TGIS* specialist and the potential *TGIS* users. The example below refers to successful applications of *TGIS*-based hypoth-

[1] Another interesting feature of knowledge synthesis is that, work done in one scientific discipline can be used to elucidate certain aspects of another. This feature usually involves the borrowing of cognitive metaphors from other disciplines in order to shed a different light on specific problems of the discipline of interest. The reader is referred to Sect. 2.3 for a brief discussion of metaphors.

[2] Epistemic analysis is concerned with the theory of knowledge (i.e., the acquisition, structure, processing, as well as the criteria of knowledge); see, also, discussion in Chap. 4 and 5.

esis testing and decision making in scientific disciplines such as agriculture, epidemiology, and ecology.

Example 1.2. Sauer (1952) used maps of agricultural hearths (tropical and seed planters, etc.) as hypotheses, and then combined them with a set of deductive logical rules in order to suggest the origins and dispersal of agriculture in the New World. Sauer's map-based approach was extended in an epidemiological setting by Haggett (2000) to study disease origins. Hilborn and Mangel (1997) describe an ecological context in which spatiotemporal maps of Mediterranean fruit fly (medfly) populations plays an instrumental role in testing different theories of medfly outbreaks in California. Such a theory testing is crucial for a number of reasons, including the understanding of the basic biology of medfly invasions, determining the implications of outbreaks on alternative agricultural practices (medfly outbreaks are some of the most destructive in agriculture, causing multi-million dollar damages per year, etc.) and, optimally, assisting the decision making process on how to prevent future outbreaks. ▪

In view of the above considerations, it becomes rather obvious that the determination of the *application environment* is important to *TGIS* specialists, for a number of reasons. The effectiveness of a *TGIS*, e.g., needs to be evaluated on the basis of its contribution to problem-solving within the particular application environment. In many situations, a deeper understanding of such an environment requires that the *TGIS* specialists work closely with the intended *TGIS* users in the relevant scientific domains. Such a collaboration allows the *TGIS* specialists to find out the information requirement of the users in order to develop a *TGIS* for supplying it. It is worth mentioning that, in certain organizational situations (related, e.g., to social policy, health management implementation, or routine marketing activities) the users can assign functional or behavioral decision criteria to the particular *TGIS*, which then can automatically make intelligent decisions and generate appropriate orders.

1.2 Field-Based *TGIS*

There is a variety of *TGIS* classifications in the literature. An important classification, which is particularly relevant to the theme of this book, is as follows (Goodchild 1989; Heuvelink 1998):

a *TGIS* dealing with *objects* which possess geometrical and topological features and non-space/time attribute values (representing points, lines, arcs, areas, etc.).
b *TGIS* dealing with *fields*, natural, social or epidemiological, which are functions taking their values in a geographotemporal domain (e.g., attribute data representing the distribution of contaminant concentration, soil erosion properties, hydrologic parameters, exposure fields, ecological patterns, and disease rates).

Although an important topic in its own right, *TGIS* of the category (*a*) above will not be considered in this book. Instead,

This book will focus on field-based TGIS, since the scientific disciplines of interest are concerned mainly with fields distributed in space and time.

Space/time maps are viewed as science-based representations of reality, and as such, they are concerned with two central elements of this reality: spatiotemporal physical geometry (coordinates and metrics), and fields (natural attributes, epidemiological magnitudes and social qualities). The representations portray only the information that the *TGIS* specialists have chosen to fit the intended uses of the map, rather than each and every detail of the actual fields, which would be neither practically possible nor desirable.

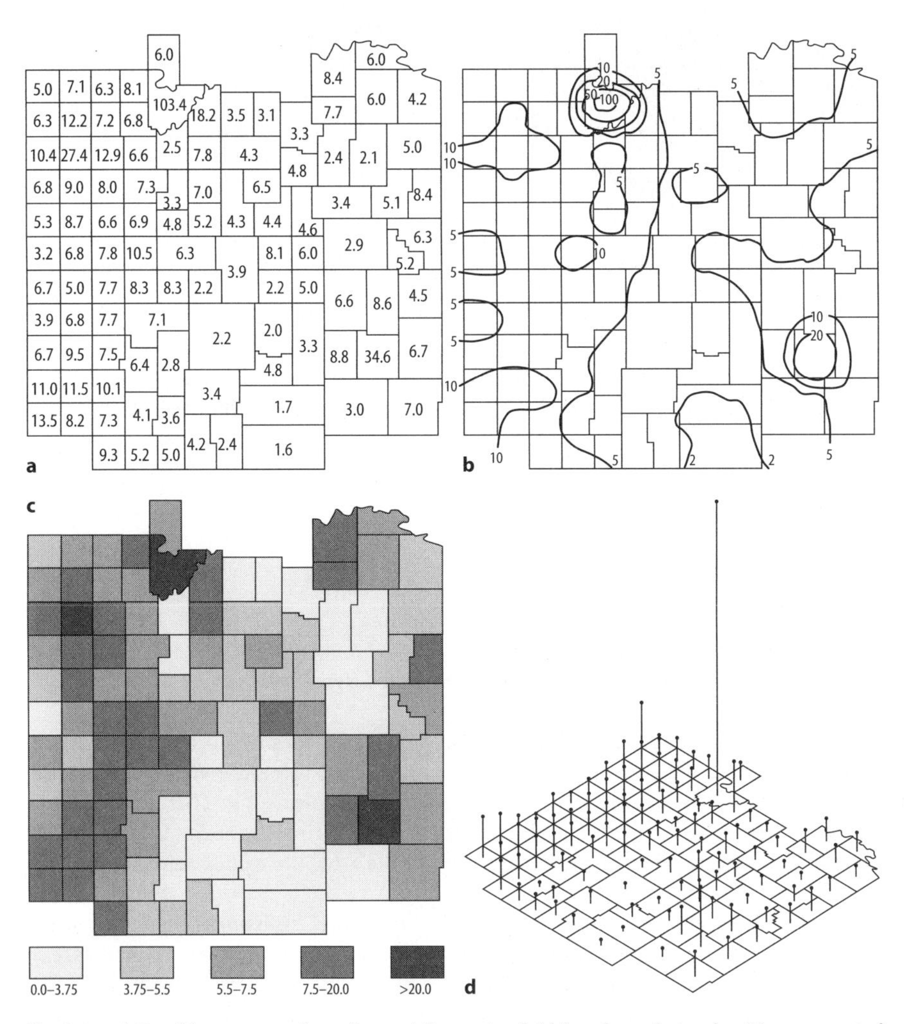

Fig. 1.1 a–d. Possible representations of a spatially varying field (rural population densities over part of Kansas). **a** As an array of population density values through space; **b** as an isocontour map; **c** as a choropleth map; **d** as a set of elevated points (modified from Jenks and Caspall 1971; Robinson et al. 1995)

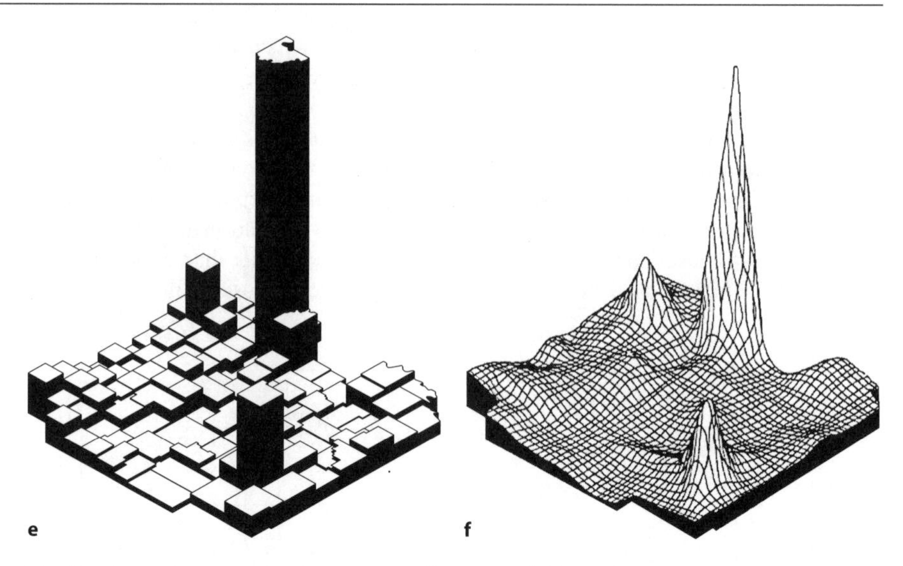

Fig. 1.1 e,f. Possible representations of a spatially varying field (rural population densities over part of Kansas). **e** As a set of erected prisms; and **f** as a smoothed surface (modified from Jenks and Caspall 1971; Robinson et al. 1995)

There are several ways to represent/visualize a field that varies across space and time. For illustration, Fig. 1.1 presents a series of possible representations of a spatially varying field of rural population densities over part of the state of Kansas. In particular, such a field can be visualized as:

a An array of attribute values assigned at the centers of representative areas.
b An iso-contour map.
c A choropleth map.
d A set of elevated points at the centers of the areas.
e A set of prisms erected over the areas.
f A smooth surface of the actual and interpolated densities.

The first set of maps in Fig. 1.1 (i.e., a, b, and c) provides two-dimensional representations, whereas the second set (i.e., d, e, and f) offers three-dimensional representations. Each set has its own special features and profitable uses in *TGIS*. For example, while the first set of maps is more appropriate for planimetric tasks, the second set gives a stereometric view of the distribution of the attribute values across space.

Example 1.3. In global warming studies the *integrated assessment modelling* (usually referred to as *IAM*) plays an important role (Houghton 1997). Integrated assessment modelling consists of various components which represent natural fields (including physical, chemical, and biological fields) governing the space/time distribution of greenhouse gases in the atmosphere, their effects on climate, sea level, and ecosystems, see Fig. 1.2. Furthermore, these components determine the climatic, ecological, human health, and socio-economic effects of varying greenhouse gas concen-

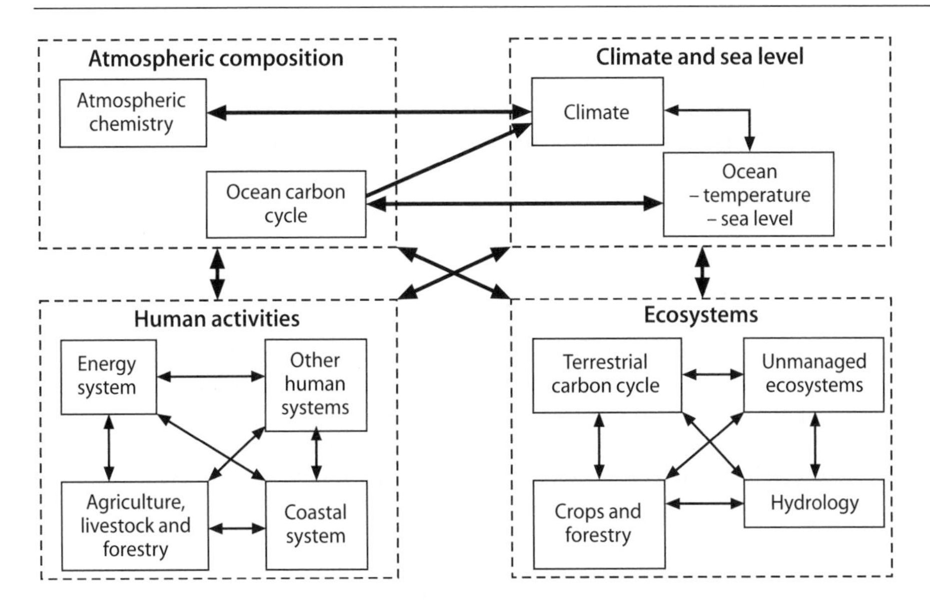

Fig. 1.2. Outline of a full scale Integrated Assessment Modelling

trations. *TGIS* are valuable and highly sophisticated instruments of integrated assessment modelling (they improve inference on the basis of nonlinear interactions between the various components, etc.). ▪

For many scientists, *TGIS* constitutes a major technological breakthrough due to its unique features, such as visualization power, considerable flexibility, and an ability to analyze a variety of data sources at different space/time scales. The *TGIS*-generated maps across space and time can be manipulated, transformed and selectively interrogated by the experts. As with any technology, however, the *TGIS* technology has its own limitations. It can be a relative expensive technology, and its results can be misinterpreted or misused by not properly trained persons (for a thorough review, see Osleeb and Kahn 1999). Additional limitations associated with the new realities in the *TGIS* area have been widely recognized in recent years. For instance, in a recent book Birkin et al. (1996) emphasize the fact that:

> Most of what is available in the current *GIS* industry is limited in use to descriptive analysis, and far too restricted in real analytical power. What is required is the suite of model-based tools … in conjunction with data storage, retrieval, and graphic facilities.

Furthermore, Lagran (1992) points out that, among the requirements that are needed to drive the development of *TGIS* are an adequate conceptual framework of spatiotemporal field variation and efficient techniques to operate on the physical space/time data. Many other authors have also recognized the need for

Theoretically sound and practically efficient methods of spatiotemporal analysis and mapping

(e.g., Arlinghaus 1995; Fisher 1995; Kraak and Ormeling 1996). The present book is an attempt to demonstrate to the *TGIS* community that *Modern Spatiotemporal Geostatistics* can address such needs of the field-based *TGIS* in a mathematically rigorous and physically meaningful manner – one that avoids the quick and easy solutions of naive empiricism[3], which can distract from dealing with the deeper problems underlying many modern *TGIS* applications.

1.3 *TGIS* Functions

Traditionally, a *TGIS* has been characterized by the set of its *functions*. Depending on the type of *TGIS* considered (field-based or object-based, etc.), various such sets have been proposed in the literature (see, e.g., Clarke 1986; Rhind and Green 1988; Chrisman 1997). The *TGIS* functions are usually classified into two major groups of so-called fundamental functions and advanced functions. Below we give a more detailed description of these two groups of functions:

1. The *fundamental functions* involve low-order geometric operations and may be viewed as tools establishing relationships between spatial and temporal entities[4]. Typically, these functions include
   i. *Measurement* functions, which enable calculations associated with points, lines, areas, and volumes (e.g., measuring the distance between two points along a straight or a curved line, measuring the areal extent or the volume of an object).
   ii. *Classification* functions, which group objects into classes according to the goals of the analysis (e.g., attribute-based classification involving comparison operations, or topology-based classification assigning new values as a function of the position, the shape, etc. of the spatiotemporal field pattern considered).
   iii. *Scalar* functions, which change the attribute values of the input layer by a given figure (e.g., adding/subtracting a specified constant to each attribute value, multiplying/dividing each attribute value by a specified constant, or exponentiating each value to a specified constant).
   iv. *Overlay* functions, which produce attribute values on an output layer by combining the attribute values on two or more input layers (e.g., arithmetic overlay functions such as addition/subtraction and multiplication/division; algebraic and statistical operations, such as averaging and ranking; logical operations such as intersection, union and complement).
   v. *Neighborhood* functions, which assign values to a space/time point on the output layer according to the features of its local neighborhood in the input layer (e.g., search procedures involving the sum, mean, or variance of the values within a specified window, the maximum/minimum value or the number of different classes in the window, etc.; surface operations which calculate topographic features such as slope and aspect; and interpolation functions which estimate unknown attribute values on the basis of the data at neighboring points).

[3] Naive empiricism and its implications in *TGIS* are discussed in Sect. 4.1, and references therein.
[4] A comprehensive overview of the existing fundamental functions is given in Malczewski (1999).

Table 1.1. Examples of *GIS* with statistical functions (modified from Malczewski 1999)

Software	Developer/URL address	Statistical functions include
ARC/INFOGRID	Environmental Systems Research Instit., Inc. www.esri.com	Descriptive statistics, regression, spatial correlation, multivariate classification, surface interpolation, trend surface analysis
IDRISI	Graduate School of Geography, Clark Univ. www.idrisi.clarku.edu	Descriptive statistics, sampling, regression, spatial correlation, multivariate classification, surface interpolation, quadrant analysis, time series analysis
SpaceStat	Anselin (1995) www.rri.wvu.edu/spacestat.htm	Exploratory spatial data analysis, Moran scatterplot outliers, scatterplot brushing
S+SpatialStats	MathSoft, Inc. www.mathsoft.com	Spatial correlation, variograms, kriging, spatial regression, trend surface analysis

vi. *Connectivity* functions, which represent topological relationships like the linkage of points or polygons to each other (vector structures) or describe the linkage between pixels (raster data). The connectivity functions include proximity, buffering, and spread operations, as well as network analysis.

2. The *advanced functions* (also called, *compound operations*) provide mathematical techniques for rigorous and efficient data processing. Most existing *GIS* use a range of statistical functions, which are designed to account for random data and spatial and temporal uncertainties of the attributes of interest (Table 1.1). These functions include the following:
   a *Classical statistics*, i.e., using conventional statistical techniques in *GIS* data processing (e.g., descriptive statistics, multivariate analysis, cluster and discriminant analysis, principal components, sampling, and factor analysis). These techniques are based on random, spatially independent fields and, thus, no consideration is given to the specific nature of the geographical data.[5]
   b *Time series analysis*, which focuses on a purely temporal analysis of data localized in space. Most time series techniques are based on the fundamental work of Wold (1938), Kolmogorov (1939), and Wiener (1949). Their use in modern geographical applications is rather limited.
   c *Classical geostatistics and spatial statistics*, which are concerned mostly with the purely spatial description of the data. The techniques used in spatial analysis are, to a considerable degree, extensions in a spatial domain of the time series techniques above. Such systematic extensions include the works of Matern (1960), Gandin (1963), and Matheron (1965).

In some cases, an attempt has been made to combine time series techniques with spatial statistics (e.g., Bennett 1979). Unfortunately, this combination was often proven to be rather artificial and inadequate, since it did not capture salient features of spatiotemporal variation (such as continuity, relative correlation structure in space

[5] These techniques violate basic laws of geography, such as Tobler's law (e.g., in geography everything is related to everything else, and near things are more related than others); see, Chrisman (1997).

and time, natural heterogeneity, and physical connections); questionable assumptions of independence were involved; in many cases the spatial fields were estimated separately from the temporal processes, etc.

In view of the modern *TGIS* needs discussed in the preceding Sect. 1.1 and 1.2, important limitations of the functions (*a*), (*b*), and (*c*) above include the following:

i. They do not provide a *composite space/time framework* for physical knowledge integration and processing; and
ii. They do not account for important sources of knowledge, such as the *natural laws* and *theoretical models* of life support sciences, in a rigorous and systematic manner.

In many real-world applications, the limitations (*i*) and (*ii*) can pose serious obstacles to a physically meaningful and technically sound *TGIS*. Several studies have shown, e.g., that composite space/time analysis can lead to more accurate and informative maps than spatial statistics or time series techniques[6]. Also, physical law-based modelling has to evolve to provide the means for sound scientific inferences rather than mere statistical inferences.

1.4 Novel Contribution to *TGIS*

In the last section of his influential book titled "Time in Geographical Information Systems", Langran (1992) describes a set of important *TGIS*-related problems that beg solution:

> Analytical problems include how to treat multi-scaled temporal data, how to interpolate and generalize between spatiotemporal samples, what types of analysis could reveal patterns in space and time, and how the past can serve in forecasting the future.

A contribution of the advanced functions discussed in the present book is that they offer mathematically rigorous and physically meaningful solutions to the above problems in the context of field-based *TGIS*. Within such a framework, a central goal of the book is to introduce *TGIS* specialists to the *Bayesian maximum entropy* (*BME*) approach of *Modern Spatiotemporal Geostatistics*[7]. In this respect, one of the main theses we would like to promote in this book is that

The contribution of the BME *approach should be viewed primarily in the context of advanced, field-based* TGIS *functions.*

In particular, an important goal of the *BME*-based *TGIS* is to address the limitations of the existing functions, as described in the previous sections.

[6] Bogaert and Christakos (1997), Christakos and Vyas (1998), Christakos and Serre (2000).
[7] In addition to the *BME* approach, the *Modern Spatiotemporal Geostatistics* paradigm has produced several other space/time modelling and mapping techniques, some of which are briefly discussed in Chap. 4 and 5. *BME*, however, includes some of the best developed (analytically and computationally) space/time techniques, at the moment. As a result, the focus of this book will be on advanced *TGIS* functions that are based on the *BME* approach.

1.4.1
BME-Based Advanced Functions

Within the above general framework, the following three *BME*-based *advanced functions* will be discussed in this book:

1. *Knowledge base* (*KB*) *development*. This function involves the acquisition and storage of the various forms of data and other knowledge sources (physical laws, ecological principles, epidemiological relationships, etc.) relevant to the problem at hand.
2. *Spatiotemporal representation*. In this function one chooses the appropriate space/time geometry, establishes the epistemic rules of *KB* integration and processing, selects the appropriate mathematical tools, and clarifies the goals of the study.
3. *Scientific mapping*. This function demonstrates the ability of *BME* to offer valuable information about the phenomena of interest in terms of composite space/time maps and probability distributions, and to aid scientific interpretation and decision-making via theory-laden visual representations.

We should notice that each one of the three *BME*-based advanced functions above may involve one or more of the traditional *TGIS* functions, while addressing some of the technical requirements of modern spatiotemporal analysis and mapping. As a matter of fact, spatiotemporal analysis and mapping in the sense considered in this book includes *data mining* techniques (see, e.g., Limb and Meggs 1995) as limiting cases, since *BME* functions go far beyond the limited scope of most data mining techniques (by incorporating a considerably larger variety of physical knowledge bases, relying on epistemically sound principles rather than on purely computational schemes, providing better models of the real world, etc.).

1.4.2
Stochastic Modelling

The three *BME*-based advanced functions are founded on the *stochastic modelling* of *KB* (natural laws, scientific theories, etc.) in an integrated space/time domain, whereas the classical statistics-based functions (see, e.g., Sect. 1.3 above) merely use formal techniques of pattern fitting (trend projection, regression analysis, sampling theory, etc.). As a consequence,

The BME*-based functions capture the main interrelations of actions-reactions and cause-effect associations in operational* TGIS *situations.*

Therefore, the functions offer valuable insights into the behavior of the life support systems being modelled, and useful assessments of the underlying space/time uncertainties and natural heterogeneities. These remarkable features of the *BME*-based advanced functions enhance their scientific content and make them a central force in the study of such diverse geographotemporal phenomena as air pollution distribution, flow and contaminant transport in environmental media, meteorological data assimilation, health damage indicators, gene frequency variability, biodiversity trends, population dynamics, and epidemiologic patterns across space and time.

Table 1.2. *TGIS* software with statistical functions

Software	Developer/URL address	Statistical functions include
BMElib	Center for the Advanced Study of the Environment (CASE), UNC-CH. www.sph.unc.edu/envr/case	Descriptive statistics, spatiotemporal covariances, variograms, and multiple-point correlations, space/time *BME* mapping, kriging, physical modelling, Bayesian and non-Bayesian conditionalization

1.4.3 *BMElib* Software

Table 1.2 gives a brief description of the *BMElib* computer software used in *TGIS* modelling and composite space/time mapping applications. The reader can find a detailed, step-by-step presentation of a version of the *BMElib* in Chap. 7. This version is appropriate for the goals of the present book. For example, the reader can use it to reproduce the results of several of the numerical experiments discussed in the book. It should not be ignored that, as is usually the case with computationally advanced formalisms, the successful application of the *BME* functions in *TGIS* practice, will depend to a considerable extent on the computational resources available (this dependence becomes increasingly valid, as more complicated sets of physical equations and site-specific data are included in the field-based *TGIS* study).

1.4.4 Epistemic Viewpoint

From an *epistemic* viewpoint, the *BME* approach offers a powerful instrument of knowledge *synthesis* (Sect. 1.1), i.e., it contributes to the adequate synthesis of a wide variety of knowledge sources.[8] At the same time, *BME* can help stimulate *spatiotemporal thinking* and visualize situations that would be otherwise invisible. Within the *BME* context,

Methodology, formal analysis, scientific theory and practice are intimately connected.

The methodology offers a perspective of reality which is based on a specific philosophical framework. The latter is dictated by ontological and epistemic ideas which have a significant effect on the *TGIS* project under consideration (i.e., the choice of research topics, methods of formal analysis, observation techniques, interpretation of the results and justification of the conclusions). The epistemic viewpoint brings to the fore the human role of the *TGIS* specialist. A specialist may select certain elements of a geographotemporal situation as being the most significant for study and representation by the *BME* formalism. This selection depends on factors which reflect the specialist's own

[8] It is true that, sometimes these sources involve competing proposals (e.g., different models representing the same physical situation). In our view, what makes the knowledge synthesis aspect of *BME* so inviting is precisely the attempt to find a balance between divergent or even opposing proposals, to draw out and combine that which is valuable in each one of them.

skills, beliefs, perceptual processes, values, motivations, and prejudices (e.g., a specialist could be prejudiced against a specific technique because of his lack of understanding or inexperience with it). To avoid undesirable complications, *TGIS* specialists should identify in advance which factors may influence their own actions and decisions.

1.4.5 Scientific Hypothesis Testing and Explanation

A relevant issue of considerable interest is whether or not the *BME*-based composite space/time maps and technical reports generated by a *TGIS* have any part to play in *scientific hypothesis testing* and *explanation*. The answer is in the affirmative. For a detailed discussion of this important issue the reader is referred to Chap. 8 of this book. Real-world examples abound in life support sciences (natural, social, epidemiological, etc.). Christakos and Kolovos (1999), e.g., combined spatiotemporal maps of ozone concentration, biological burden, and health effects to derive distributions of expected population damage indicators due to pollutant exposure in parts of the Eastern USA. Christakos and Serre (2000) used *BME*-based maps of temperature and mortality throughout the state of North Carolina to test hypotheses concerning population exposure-health effect associations. Furthermore, Choi et al. (2001) analyzed California mortality data at different scales (e.g., county vs. zip-code scales), thus obtaining a quantitative assessment of the effect of the scale hypothesis on the mortality predictions across space and time.

1.4.6 Revisionistic Paradigm

To put things in a methodological perspective, a diversion may be necessary at this point. *Mutatis mutandis*, many *TGIS* experts view *Modern Spatiotemporal Geostatistics* as a shift – in a Kuhnian sense – from the old reigning paradigms of classical geostatistics and spatial statistics to a new, kind of a *revisionistic*, paradigm[9]. Regardless of how useful a reigning paradigm has been proven in the past, the time comes that its limitations pile up and it must be replaced by a new paradigm (see, e.g., Kuhn 1962). Nevertheless, the resistance to change often demonstrated by the practitioners of the reigning paradigm has been well-documented by historians and philosophers of science. Indeed, as Klee (1997) emphasizes there are certain institutional reasons why the reigning paradigm is inherently "conservative" in the face of any new ideas:

> No practitioner who has built his career out of work that depends on the correctness of a particular paradigm wants to see that career's worth of work questioned under some different paradigm that might very well render the research worthless, or at best, require that the work be repeated under slightly different conditions of evaluation.

A product of the revisionistic paradigm proposed by *Modern Spatiotemporal Geostatistics* is the *BME* space/time modelling and mapping approach. The new paradigm replaces certain fundamental principles of the old paradigm (purely statistical criteria, physical theory-free techniques, naive inductivism, etc.) with epistemically sound

[9] The term "paradigm" is used to describe a particular way of looking at things. The paradigm includes a set of theories, techniques, applications, and instrumentation together (see, also, discussion in Sect. 8.1 and 8.5).

principles (which go far beyond pure inductivism into the realm of hypothetico-deduction, emphasizing scientific hypothesis testing, explanation and prediction). Furthermore, the *BME*-based techniques account for important bases of general and site-specific knowledge (in the form of physical laws, scientific theories, phenomenological relationships, hard data, various types of soft or uncertain information, etc.) in a mathematically rigorous and methodologically systematic fashion. *Modern Spatiotemporal Geostatistics*'s effort to repair anomalies of the reigning paradigm and provide a new framework of scientific analysis and modelling has been met with the typical reactions anticipated by the Kuhnian schema of scientific methodology. Reigning paradigms always die hard, because there is too much at stake (egos, research funding, social recognition, etc.). It is a fact of life that practitioners of the reigning paradigm rarely acknowledge its inadequacies. Instead, they usually make an "after the event" effort to convince each other and the revisionists that there is no major problem or that the reigning paradigm can take care of the anomalies. According to Klee, allegiance to the reigning paradigm affects everything the practitioner does in the course of research. It even alters the way the practitioner "sees" the observational data. In fact, the practitioner sees the data a certain way because prior training in the paradigm has "cut off" from cognitive access all alternative formulations that do not fit the paradigm. In their rush to show that "things are under control", the practitioners of the reigning paradigm suggest modifications which supposedly overcome some of the limitations pointed out by the revisionists. More often than not, this is an effort whose only success is to demonstrate further the inadequacies of the old paradigm. As Kuhn (1962) famously pointed out, the sudden appearance of various alternative and often incompatible versions of a paradigm that previously reigned in monolithic majesty is a sure sign that the scientific discipline has gone into crisis and that the comforts of the formerly reigning paradigm have ended for practitioners in that discipline.

1.5 Concluding Remarks

There is no doubt that *TGIS* are very valuable tools in the scientific study of many natural phenomena, social processes and epidemiological fields. Of course, *TGIS* have not been developed to the same degree of sophistication in all scientific disciplines. The physical applications of *TGIS*, e.g., are way ahead of many of its epidemiological and biostatistical applications. This is unfortunate, because

Very few quantitative tools are as valuable to epidemiological investigations across space and time as is the TGIS *technology.*

In our view, it is necessary that some disciplines overcome any obstacles generated by professional correctness[10] and develop interdisciplinary links with natural sciences so that they can learn from their extensive experience with *TGIS*.

[10] According to professional correctness, interdisciplinarity is often considered professionally suicidal (e.g., Michael 2000). The very existence of the discipline or profession itself depends on its apparent autonomy, in the sense of having the primary responsibility for doing a job the society wants done. Unfortunately, professional correctness often causes scientific progress to take a back seat.

TGIS is a dynamic technology that continuously undergoes through various stages of review and development so that it can respond adequately to the increasingly more complex needs of interdisciplinary applications. As a matter of fact, the existing *TGIS* advanced functions need to improve in order to provide a realistic geographotemporal representation (which is usually characterized by a heterogeneous data base, uncertain information sources, etc.) with

Spatiotemporal continuity and internal physical consistency.

With this goal in mind, the advanced functions discussed in this book seek to offer

Powerful logical concepts and sophisticated mathematical techniques

as well as a

Practical guide to the new realities of TGIS

by promoting the idea of "strategic flexibility" in *TGIS*, i.e., being able to set a clear general direction yet also being capable of adapting as the requirements of the problem change and new opportunities arise. The *BME* approach involves both, modelling and analyzing *TGIS*. In particular, modelling is the creative side of *TGIS*, and analyzing is the critical side.

We conclude this introductory chapter by amending the famous Cartesian dictum "I exist, therefore I am", with the following:

I think, therefore I bme.

It is our hope that the readers will share our optimism or, at least, they will enjoy the humor.

Chapter 2

Spatiotemporal Modelling

Sometimes scientific ideas, like strange
musical compositions of surrealistic dramas,
need a ready audience as well as a creator.
D. Lindley

2.1 Spatiotemporal Continuum

The historical role of geometry in geographical mapping and cartography is properly emphasized in Robinson et al. (1995):

> A big change in mapping occurred following the development of geometrical concepts by Greek scholars. Geometry provided the means for determining the shape and size of the Earth and for determining the relative position of environmental features. Geometrical concepts also provided a foundation for the development of location reference systems...

Nowadays, geographotemporal decision-making in life support sciences often requires the use of datasets where, in addition to the measured values, the description of the physical geometry (in terms of spatial and temporal coordinates) is also part of these sets. In the case of a contaminated site, e.g., making decisions about a suitable cleanup strategy will depend heavily on the spatiotemporal characteristics of the samples obtained. In medical geography, contagious spread of diseases is a process which occurs during a specified time period and is strongly influenced by spatial distances, since nearby populations have higher probability of contact than remotely located populations. Also, an epidemiologic survey presented in a tabular form without any reference to spatial locations and times may prevent the adequate interpretation of the survey[1].

As a consequence of the above considerations, the scientific analysis of similar geographotemporal situations requires the introduction of

a The notion of a *spatiotemporal continuum* equipped with a coordinate system and a measure of space/time distance.
b Models and techniques that establish *linkages* between spatiotemporally distributed data.

A spatiotemporal continuum (or domain) $\mathcal{E}$ is a set of points associated with a continuous spatial arrangement of events combined with their temporal order. Spatiotemporal continuity implies an integration of space with time and is a fundamental property of the mathematical formalism of natural phenomena. The events represent attributes related to a natural phenomenon or field (e.g., contaminant concentration, hydraulic head, temperature, or fluid velocity). Within the spatiotemporal continuum $\mathcal{E}$, space represents the order of coexistence of events, and time represents the order of their succes-

[1] Mausner and Kramer (1985) maintained that "Whatever the objective of study, the concept of place, i.e., a clearly defined geographic area with its related population, is central to epidemiological study". Hagget et al. (1977) stress the need to use locational models.

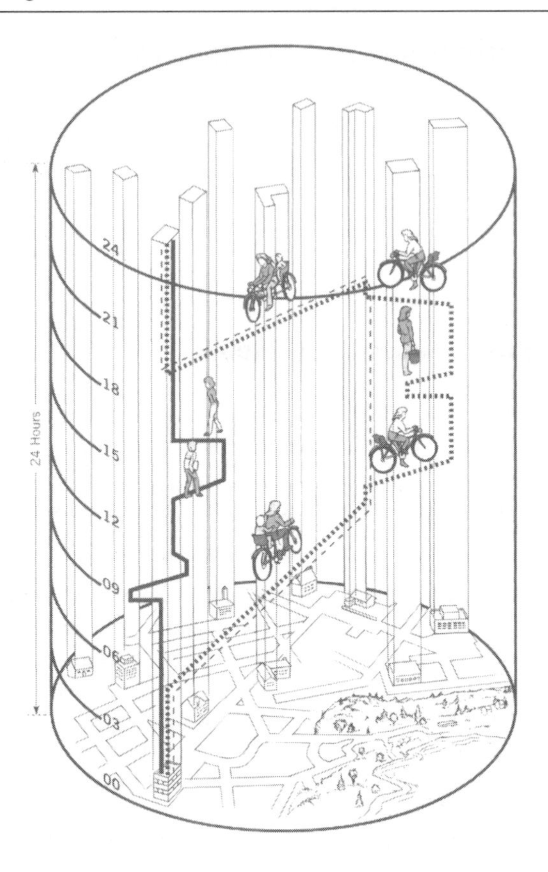

Fig. 2.1.
Spatiotemporal diagram describing typical life histories of common people across space and time (from Chrisman 1997; ©1997 by J. Wiley & Sons, Inc., reproduced with permission)

sive existence. In human exposure studies, e.g., a possible representation of $\mathcal{E}$ is given in Fig. 2.1. This figure presents a space/time diagram of typical life histories (series of events taking place in space and time, etc.) of common people during their daily activities. Changes in the spatial distribution of human populations over time can have a substantial effect in the study of certain health conditions (e.g., diseases with long latency periods), the epidemiological analysis of spatial clusters (e.g., due to human mobility, what was a disease cluster at a certain time period may be not so at a later period), etc.

For the continuum $\mathcal{E}$ to be useful in real-world *TGIS* applications it must be equipped with a *coordinate system* identifying points in space/time and a metric that measures "distances" in space/time[2], i.e.,

$$\mathcal{E}\text{: (coordinates } \boldsymbol{p}\text{, metric } |d\boldsymbol{p}|)$$

Generally speaking, a coordinate system is a systematic way of referring to places, times, things and events. A point in a spatiotemporal domain $\mathcal{E}$ can be identified by means of

[2] A detailed discussion of the geometrical properties of a spatiotemporal continuum can be found in Christakos (2000). In this chapter we provide a rather brief exposition which is sufficient for the goals of the present book.

two separate entities: the spatial coordinates $s = (s_1, \ldots, s_n) \in S \subset R^n$ and the temporal coordinate t along the temporal axis $T \subset R^1$, so that the combined space/time coordinates are

$$p = (s,t) \in E = S \times T \quad (2.1)$$

The above union of spatial coordinates s and the time coordinate t is defined in terms of their Cartesian product $S \times T$. Equation 2.1 suggests several ways to "locate" a point in space/time. Essentially, the only constraint on the coordinate system implied by Eq. 2.1 is that it possesses n independent quantities available for denoting spatial position and one quantity for denoting time instant (one can use this approach, e.g., in the case of Fig. 2.1). An interesting classification of the spatial coordinate systems (s) can be made in terms of the following two major groups:

- The *Euclidean* group of coordinate systems, which includes systems for which there exists a transformation to *rectangular* (*Cartesian*) coordinates ($s_1, \ldots, s_n$; $n = 2$ or 3).
- The *non-Euclidean* group of coordinate systems, which includes systems for which it is not possible to perform a transformation to Cartesian coordinates.

In the two-dimensional case ($n = 2$), e.g., the former group is associated with a flat, Euclidean geometry (coordinate systems on a Euclidean plane can be transformed to a rectangular system), whereas the latter group is associated with a curved, non-Euclidean geometry (rectangular coordinates do not exist on a non-Euclidean surface). Table 2.1 gives a summary of some commonly used Euclidean coordinate systems, which belong to the group of orthogonal curvilinear coordinate systems (the polar, cylindrical and spherical systems can all be transformed to a rectangular system). Cartesian coordinate systems used by U.S. agencies include the universal transverse mercator (UTM) grid system, the universal polar stereographic (UPS) grid system, the state plane coordinate (SPC) system, and the public land survey (PLS) system (Wolf and Brinker 1989). Basic non-Euclidean coordinate systems are the *Gaussian* coordinate system, and the *Riemannian* coordinate system. Other coordinate systems are mentioned in Table 2.2[3]. These are systems of coordinates with certain particular physical properties (e.g., cyclidic coordinates are such that the Laplace equation is separable; toroidal coordinates are such that the equation of a magnetic-field line is that of a straight line in these coordinates). Therefore, we come to the following conclusion:

TGIS modelling must be associated with a real-world space/time coordinate system that is physically appropriate for the situation at hand.

Table 2.1.
Common Euclidean coordinate systems

Rectangular	Polar	Cylindrical	Spherical
s_1	r	r	ρ
s_2	θ	θ	φ
s_3	–	s_3	θ

[3] For more details, the interested reader is referred to Weisstein (1999); and references therein.

Table 2.2. A partial list of coordinate systems with useful physical properties

Barycentric	Chow	Ellipsoidal	Hamada	Quadriplanar
Bipolar	Conical	Glebsch	Orthocentric	Toroidal
Bispherical	Cyclidic	Grassmann	Parabolic	Trilinear

The latitude/longitude coordinate system, e.g., presupposes a specific model of the Earth. In practice, one may need to establish a transformation of the original coordinate system into one that offers the most realistic representation of the physical fields involved.

Example 2.1. Traditionally, *geographic coordinates* are expressed in terms of latitude and longitude (Fig. 2.2). A line on the surface of the Earth that joins the North and South poles is a meridian. The *latitude* of a point P on the surface of the Earth is defined as the angle ϕ_P between P and the equator along the meridian. The *longitude* is defined as the angle θ_p between the meridian through P and the central meridian (through Greenwich, U.K.) in the plane of the equator. ▪

In *TGIS*, the process of registering data to a coordinate system such as the one discussed above is usually termed *georeferencing*. Georeference systems are not always stable, but they may change with time. A powerful technology for precise determination of coordinate positions is the *Global Positioning System* (*GPS*). In recent years, due to its ability to provide immediate and highly accurate coordinate positions, *GPS* has become a major technology for *TGIS* (e.g., Davis 1996).

The second feature of a spatiotemporal continuum $\mathcal{E}$ is its *metric*, that is, a mathematical expression that defines distances in space/time. A metric $d\boldsymbol{p}$ is a function defined for a coordinate system such that the spatiotemporal distance between any two nearby points in that system is determined from their coordinates. In *TGIS*, it is usually convenient to consider two prime metrics: one is the separate metric and the other is the composite metric. Let us consider, first, the separate metric.

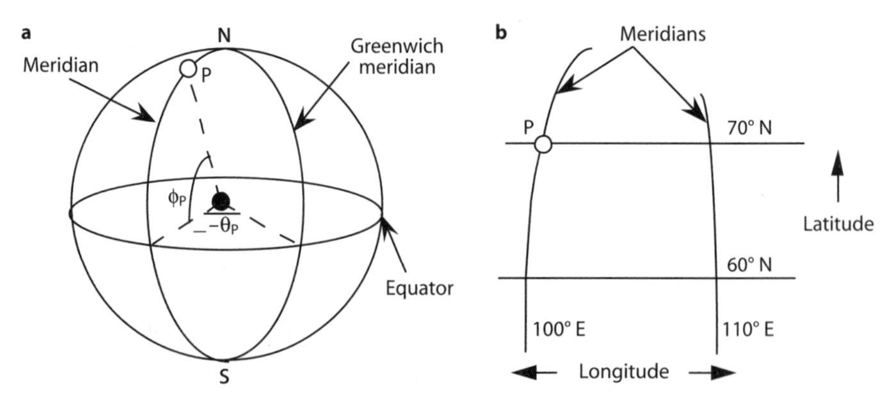

Fig. 2.2. a Geographic coordinates of *P*; ϕ_P = latitude and θ_p = longitude. **b** An example where *P* (70° N, 100° E)

- The *separate* metric includes a spatial distance $|ds|$ and a time interval dt,

$$dp: \; (|ds|, dt) \in R^1_{+,0} \times T \tag{2.2}$$

where $R^1_{+,0}$ is the non-negative part of R^1. These metrics are convenient for many *TGIS* applications, because they treat the concept of distance in space and time separately. Below we present a few spatial distances in the case of a separate metric.

Example 2.2. Traditionally, the separate space/time geometry (Eq. 2.2) is equipped with the spatial *Euclidean* metric

$$|ds|_e = \sqrt{\sum_{i=1}^{n} ds_i^2} \tag{2.3}$$

In many other cases one assumes a non-Euclidean spatial metric of the *absolute* form (also called, the *1-norm* form)

$$|ds|_a = \sum_{i=1}^{n} |ds_i| \tag{2.4}$$

There exist physical applications in which a *fractal* metric of the form

$$|ds|_f = [|ds|_e]^{\alpha} \tag{2.5}$$

where α is a real coefficient, represents spatial distances more adequately. ▪

Example 2.3. Geographic distance between points P_1 and P_2 on the surface of the Earth is not the straight dashed line but the curved line on its surface. This distance can be expressed in geographic coordinates as

$$|ds|_{P_1P_2} = r \arccos[\sin(\phi_{P_1})\sin(\phi_{P_2}) + \cos(\phi_{P_1})\cos(\phi_{P_2})\sin(\theta_{P_1} - \theta_{P_2})] \tag{2.6}$$

Depending on the physical, economical, etc. characteristics of the situation, the distance may be not purely geographic. In porous media studies, the *ds* of Eq. 2.2 may express the length of the maximum permeability path, i.e.,

$$|ds|_{pm} = \max_{|ds|} PM \tag{2.7}$$

were *PM* denotes a porous media permeability function. In the case of the porous medium of Fig. 2.3a, the arrows show the path of maximum permeability between the points P_1 and P_2; then, the distance between these two points is the length of the path. Also, the metric $|ds|$ may denote the minimum cost path over a terrain divided into travel cost zones,

$$|ds|_{mc} = \min_{|ds|} CF \tag{2.8}$$

where *CF* is a cost function that depends on the distance as well as on other factors (Laurini and Thompson 1995). In the situation displayed in the Fig. 2.3b, e.g., due to the high travel cost areas (shown as shaded regions) it is necessary to travel from point P_1 to point P_2 not along the dashed line but rather along the path shown with arrows, in which case the distance is the length of this path. ▪

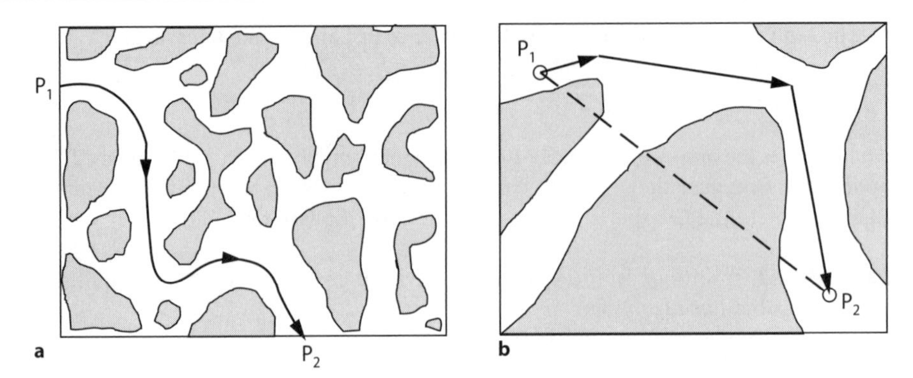

Fig. 2.3. a Path of maximum permeability distance in a porous medium. **b** Path of minimum cost distance over a terrain (the *shaded region* denotes a high-cost area)

While an Euclidean metric must be of the form of Eq. 2.3, many useful non-Euclidean metrics can be put in the general form

$$|ds| = \sqrt{\sum_{i,j=1}^{n} g_{ij} ds_i ds_j} \tag{2.9}$$

where the coefficients g_{ij} $(i, j = 1, \ldots, n)$ are functions of the spatial location. However, the fact that the g_{ij} vary in space does not necessarily imply that we are dealing with a non-Euclidean geometry. As the following example demonstrates, in some cases the dependence of g_{ij} on space is a coordinate effect rather than the result of a non-Euclidean metrical structure.

Example 2.4. In terms of spherical coordinates in R^3 the distance can be expressed as follows,

$$|ds| = \sqrt{d\rho^2 + \rho^2[d\phi^2 + (\sin^2\phi)d\theta^2} \tag{2.10}$$

Using a transformation to rectangular coordinates, Eq. 2.10 can be reduced to the Euclidean form (Eq. 2.3). Thus, while we have g_{ij} that are functions of ρ and ϕ, we are dealing with Euclidean geometry (the dependence of g_{ij} on ρ and ϕ is purely a coordinate effect). Similar transformations can be performed in the case of the polar and cylindrical coordinates of Table 2.1. ▪

What characterizes a non-Euclidean metric is the fact that one cannot derive a coordinate transformation such that Eq. 2.9 can be put into the form 2.3. This is illustrated in the following example.

Example 2.5. Spherical trigonometry provides us with the necessary rules to measure and relate the angles, sides, etc. of triangles on the surface of the Earth (considered as a sphere with radius r). The rules of Euclidean geometry do not apply to these measurements. Then, another formula expressing the distance between two points on the surface of the Earth is as follows

$$|ds| = r\sqrt{d\phi^2 + (\cos^2\phi)d\theta^2} \tag{2.11}$$

where now $d\phi$ and $d\theta$ are the latitude difference and longitude difference, respectively (both in radians). Note that the g_{ij} coefficients in Eq. 2.11 are not constants, and there is no transformation of the coordinates that will put Eq. 2.11 into the form 2.3, since Cartesian coordinates do not exist on a curved non-Euclidean surface. ▪

Comment 2.1. A remarkable consequence of the preceding analysis is that, while in certain circumstances Eq. 2.9 can be used to express a metric with constant coefficients g_{ij} in the Euclidean space (where the curvature is zero everywhere), in a (non-Euclidean) Riemannian space with curvature varying from place to place, it is not possible to transform Eq. 2.9 into some Cartesian system so that Eq. 2.3 applies and the g_{ij} are spatially invariant. The Riemannian space replaces the orthogonal transformation with a more general set of transformations which has the property that Eq. 2.9 remains invariant, and geometric quantities such as scalars, vectors and tensors transform in a certain manner under this set of transformations.

- The *composite* metrical structure assumes that the space and time parameters are connected by means of an analytical expression g, i.e.,

$$dp\colon\ |dp| = g(ds_1, \ldots, ds_n, dt) \tag{2.12}$$

where g is determined from the knowledge available (topography, physical laws, etc.).

Clearly, the composite metric assumes a higher level of physical understanding of space/time, which may involve theoretical and empirical facts of life support sciences (natural, epidemiological, etc.). In certain situations this understanding suggests adopting a relational, not an absolute, theory of space/time, according to which space and time are not self-existing objects but, instead, space/time is a network of relations among distinct changing things. A special case of the composite space/time metric Eq. 2.12 is the spatiotemporal *Riemannian* metric

$$|dp| = \sqrt{\sum_{i,j=1}^{n} g_{ij} ds_i ds_j + g_{00} dt^2 + 2dt \sum_{i=1}^{n} g_{0i} ds_i} \tag{2.13}$$

where the coefficients g_{ij} $(i, j = 1, \ldots, n)$ are functions of the local coordinates (spatial location and time) to be determined from the physical *KB* available[4].

Example 2.6. In some physical applications a useful metric is given by

$$|dp| = \sqrt{ds^2 + v^2 dt^2} \tag{2.14}$$

which is obtained from Eq. 2.13 for $g_{ii} = 1$, $g_{ij} = 0$ $(i \neq j)$, $g_{0i} = 0$ and $g_{00} = v$ (v is a known parameter); note that in Eq. 2.14, $ds^2 = \sum_{i=1}^{n} s_i^2$. ▪

[4] It should be noticed that Eq. 2.13 basically refers to infinitesimal distances. In some practical applications (e.g., space/time correlation analysis and estimation) finite distances may have the same form as Eq. 2.13, but the corresponding metric coefficients are functions of the spatial and temporal distances and do not necessarily coincide with the g_{ij} above.

By way of a summary, the choice of the most adequate space/time coordinate system and metric to describe the physical world depends – to a large extent – on the structure of the fields (natural, social, epidemiological, etc.) being described. Finding a space/time domain that works always depends on our understanding of the basic features of the problem, at least at an intuitive or phenomenological level. The coordinate system and the metric used to describe the space/time continuum $\mathcal{E}$ may be independent of each other, in the sense that one may combine a Euclidean coordinate system with a non-Euclidean metric. The purpose of the previous examples was to make the reader aware of the possible pitfalls in using non-appropriate space/time coordinate systems and metrics in the physical *TGIS* context[5].

2.2 The Random Field Model

The *spatiotemporal random field* (*S* / *TRF*) model is a mathematical construction that rigorously represents the distribution of natural, social or epidemiological fields across space and time. The *S* / *TRF* model provides scientifically meaningful representation, explanation, and prediction of natural phenomena or epidemiological patterns in *uncertain* environments. These uncertainties may be due to measurement errors, heterogeneous data bases, erratic fluctuations in the space/time variation of the underlying processes, and insufficient knowledge[6].

In this section we provide a brief exposition of the *S* / *TRF* model, which is a fundamental component of the field-based *TGIS* viewpoint considered in this book. A detailed presentation of the mathematical *S* / *TRF* theory may be found in Christakos (1992) and in Christakos and Hristopulos (1998). Herein, large and small Roman characters will denote random fields and random variables, respectively; Greek characters will denote realizations (data values, etc.). An *S* / *TRF*

$$X(\boldsymbol{p}) = \{\boldsymbol{p}, \chi\} \tag{2.15}$$

is a collection of possible realizations χ for the distribution of the field at space/time points $\boldsymbol{p}$. The *S* / *TRF* can be viewed as a collection of correlated random variables

$$\boldsymbol{x}_{\text{map}} = (x_1, \ldots, x_\nu) \tag{2.16}$$

at the space/time points

$$\boldsymbol{p}_{\text{map}} = (\boldsymbol{p}_1, \ldots, \boldsymbol{p}_\nu) \tag{2.17}$$

---

[5] As we already mentioned, a systematic mathematical presentation of the space/time coordinate systems and metrical structures is beyond the scope of this book, but it can be found in the relevant literature.

[6] In the case of global warming studies, e.g., a considerable amount of uncertainty is due to our insufficient understanding of sources and sinks of greenhouse gases, clouds, oceans, and polar ice sheets (Houghton 1997).

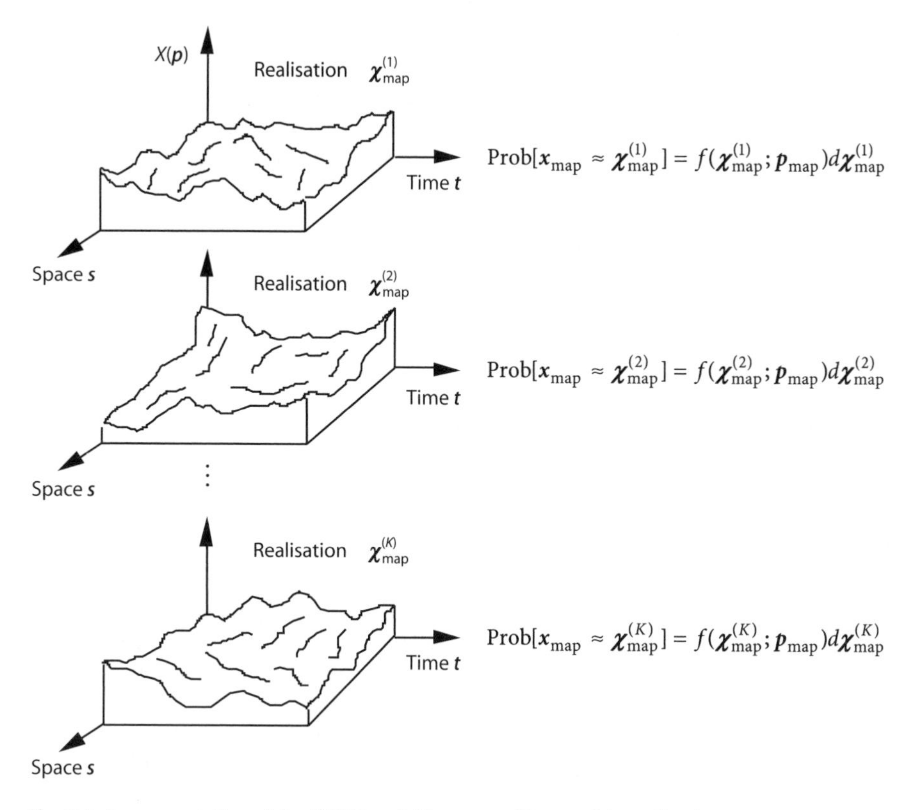

Fig. 2.4. A representation of the *S/TRF* model in terms of its possible realizations

It is noteworthy that it is not each isolated variable by itself, but its relation to the other variables that is of interest in the *S / TRF* model. A realization of the *S / TRF* at these points is denoted by the vector

$$\chi_{\text{map}} = (\chi_1, \ldots, \chi_v) \tag{2.18}$$

Then, an alternative definition of the *S / TRF*, $X(\boldsymbol{p})$, is given in terms of all possible realizations. This definition is illustrated in Fig. 2.4 by means of the 1st, 2nd and Kth realizations of the *S / TRF*. The subscript "map" in the symbols $\boldsymbol{x}_{\text{map}}$, $\boldsymbol{\chi}_{\text{map}}$ and $\boldsymbol{p}_{\text{map}}$ is used because *TGIS* seek to derive spatiotemporal maps of the phenomenon of interest.

Example 2.7. Diseases spread within a specific geographotemporal context. The diffusion processes (e.g., expansion, relocation, and combined diffusion) characterizing the geographical patterns of epidemic diseases (Haggett 2000) are typical cases of spatiotemporally varying phenomena. These processes can be represented in terms of *S / TRF* models which account for the dynamic structure of a disease spreading from one place to another, as well as for uncertain fluctuations and random patterns occurring during this spreading at one or several scales (the issue of space/time scales is an important one in epidemiology, as discussed in Sect. 4.2.4). ▪

Example 2.8. *S / TRF* models can represent *scalar* natural fields (e.g., contaminant concentration or temperature), *vector* fields (e.g., fluid velocity), or *tensor* fields (e.g., fluid stress or hydraulic conductivity in anisotropic porous formations). In the scalar case the field is represented by a single value (magnitude) at each space/time point, whereas in the vector case the field is represented by its magnitude as well as its direction at each space/time point. Tensor representations are not particularly pictorial at all. ▪

A complete stochastic characterization of an *S / TRF* is provided by the multivariate *probability density function (pdf) f* defined such that

$$\text{Prob}[\Lambda_{i=1}^{\nu}(\chi_i \leq x_i \leq \chi_i + d\chi_i)] = f(\boldsymbol{\chi}_{\text{map}}; \boldsymbol{p}_{\text{map}}) d\boldsymbol{\chi}_{\text{map}} \tag{2.19}$$

where $\Lambda_{i=1}^{\nu}$ means extended conjunction, and $d\boldsymbol{\chi}_{\text{map}}$ denotes a realization interval (note that the pdf's unit is probability per realization unit). From Eq. 2.19 scientists and engineers find it useful to derive the heuristic,

$$\text{Prob}[\boldsymbol{x}_{\text{map}} \approx \boldsymbol{\chi}_{\text{map}}] = f(\boldsymbol{\chi}_{\text{map}}; \boldsymbol{p}_{\text{map}}) d\boldsymbol{\chi}_{\text{map}} \tag{2.20}$$

which turns out to be a realistic concept given that in practice the physical attributes are measured only to some desired degree of accuracy[7]. In Fig. 2.4 we show the probability of occurrence of each realization according to Eq. 2.20. Furthermore, Fig. 2.5 presents the *hierarchy* of pdf, in which the lower the level of the hierarchy, the higher the knowledge provided by the associated *S / TRF* characterization.

A usually incomplete, yet in many practical *TGIS* applications, satisfactory characterization of the *S / TRF* is provided by a limited set of *statistical space/time moments* defined as

$$\overline{g(\boldsymbol{x}_{\text{map}})} = \overline{g}(\boldsymbol{p}_{\text{map}}) = \int d\boldsymbol{\chi}_{\text{map}}\, g(\boldsymbol{\chi}_{\text{map}}) f(\boldsymbol{\chi}_{\text{map}}; \boldsymbol{p}_{\text{map}}) \tag{2.21}$$

where $g(\cdot)$ is some known function. The reader may notice the difference between $g(\boldsymbol{\chi}_{\text{map}})$, which is a function of the realization values, and its expectation $\overline{g}(\boldsymbol{p}_{\text{map}})$, which is a function of the space/time points. In practice, usually only a limited number of space/time statistical moments is available, in which case the *S / TRF* characterization in terms of Eq. 2.21 is considered "incomplete" (in the sense that several random fields may exist that share the same moments). In the case of a multivariate Gaussian *S / TRF*, under certain conditions the multivariate pdf may be expressed in terms of the 1st and 2nd moments[8].

Fig. 2.5.
The pdf hierarchy

$$\begin{gathered} f(\boldsymbol{\chi}; \boldsymbol{p}) \\ f(\chi_1, \chi_2; p_1, p_2) \\ f(\chi_1, \chi_2, \chi_3; p_1, p_2, p_3) \\ \vdots \\ f(\chi_1, \chi_2, \ldots, \chi_\nu; p_1, p_2, \ldots, p_\nu) \end{gathered}$$

[7] For example, in a physical experiment the event $\chi = 9.36523$ is a rounding off approximation, where the actual event is $9.3652 \leq \chi \leq 9.3654$.

[8] For more technical details on the space/time moments, see Sect. 3.2.1 later in this volume.

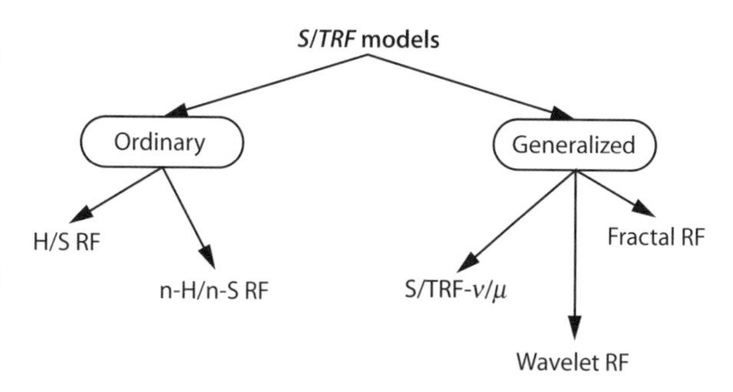

Fig. 2.6. Classification of *S / TRF* models on the basis of space/time heterogeneity and physical localization. *H / S* = spatially homogeneous/temporally stationary fields, *nH / nS* = non-*H / S* fields, ν / μ = parameters characterizing a class of *S / TRF* (from Christakos 2000)

Comment 2.2. In theory, the *g*'s can represent any function of χ_{map}. In practice, certain restrictions may be imposed by the fact that it may be difficult to calculate the corresponding moment $\bar{g}$ from the *KB* available.

There is a rich class of *S / TRF* models that can be used in *TGIS* applications. In Fig. 2.6, a model *classification* is displayed on the basis of natural space/time heterogeneity conditions (homogeneous/stationary vs. non-homogeneous/non-stationary patterns, space/time trends, etc.) and physical localization conditions (local smoothness properties, etc.). Other classifications – based on different criteria – are also possible. In practice, the implementation of one specific *S / TRF* model over another depends on the form of the space/time object variations and natural heterogeneities considered. It may, also, depend on the correspondence rules that join non-observable terms with observational terms. These rules provide the means to calculate the statistics of non-observable quantities involved in a theoretical law from the statistics of the observable quantities of an empirical law. The functional form of these statistics influences the choice of the *S / TRF* model to represent the phenomenon.

2.3 The Role of Metaphors in *TGIS*

Metaphors have been used to extend scientific theories into new domains. Generally, the purpose of a metaphor is to probe and conceptualize unknown or little understood domains by means of more familiar quantities. *TGIS* make use of certain metaphors to conceptualize important physical entities and gain valuable insight. The spatiotemporal continuum metaphor, e.g., conceptualized time as a fourth dimension, which made it possible to use the mathematics of Euclidean and Riemannian geometries. The field metaphor associates mathematical entities (scalars, vectors, or tensors) with sets of values of the natural or epidemiological processes at the space/time points. According to the complementarity metaphor, uncertainty manifested itself as an ensemble of possible field realizations that are in agreement with what is known about the phenomenon. Hence, it seems plausible to suggest that,

Mathematization of the conceptual TGIS metaphors leads to useful models.

For example, putting the above three metaphors together, and translating them into a mathematical language we got the powerful *S / TRF* model. The underlying metaphorical structure allows mathematical constructs to be linked to natural phenomena and to be regarded as scientific theories. On the other hand, the space/time predictions obtained from these theories are nonmetaphorical, since they can be verified or falsified. In this sense, the *S / TRF* model can be seen as a metaphorical system for making predictions and plotting maps of the natural distribution that can be tested empirically. Depending on one's conceptual system and its ability to function optimally in a given environment, different metaphors can be associated with a particular phenomenon which, in turn, can lead to different mathematical models of the situation. The real world can only be observable or describable in terms of such models on different *levels* (see also Chap. 8).

2.4 The Importance of Physical Geometry

The space/time metrics as defined in Sect. 2.1 above play a very significant role in the determination of the *S / TRF* characteristics (such as spatial isotropy), and the statistical moments of Eq. 2.21. Their role is also crucial in space/time predictive mapping. Symbolically, one may express these properties of the metric as in Fig. 2.7.[9]

Next, we will try to illustrate the above considerations with the help of a few instructive examples.

Example 2.9. The *TGIS* specialists, say Drs. T and R, use two different geometries to describe the same dataset. Dr. T (who is a traditionalist) uses a separate space/time geometry equipped with the usual spatial Euclidean metric (Eq. 2.3). On the other hand, on the basis of a careful study of the physical situation Dr. R (who is a reformist) concludes that a non-Euclidean spatial metric of the absolute (or 1-norm) form of Eq. 2.4 is more appropriate for the physical situation. These two choices have considerable consequences in the *TGIS* analysis of the same phenomenon. Let us discuss a few of these consequences. Basic geometric properties, like random field isotropy, acquire a different meaning for each one of the above two metrics. Consider, for instance, an isotropic random field in two dimensions (i.e., $n = 2$). While for Dr. T the spatial iso-covariance contours are circles, for Dr. R these contours are squares (see the plots in Fig. 2.8a). Furthermore, Dr. R's metric depends on the directions of the s_1 and s_2 axes, whereas Dr. J's metric is invariant under all rotations of the axes. Furthermore, for any two field points $\boldsymbol{s}$ and $\boldsymbol{s}'$, a set of points $\boldsymbol{u}$ can be defined which satisfies the betweenness norm relationship, $\|\boldsymbol{s} - \boldsymbol{u}\| + \|\boldsymbol{s}' - \boldsymbol{u}\| = \|\boldsymbol{s} - \boldsymbol{s}'\|$. In the case of Dr. T the

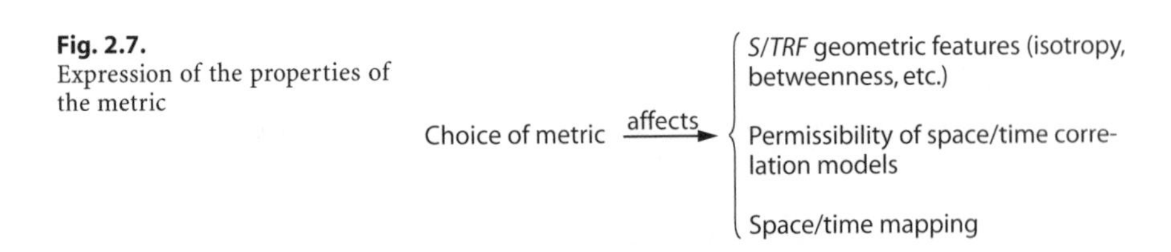

Fig. 2.7. Expression of the properties of the metric

[9] A detailed discussion of the relevant issues can be found in Christakos (2000).

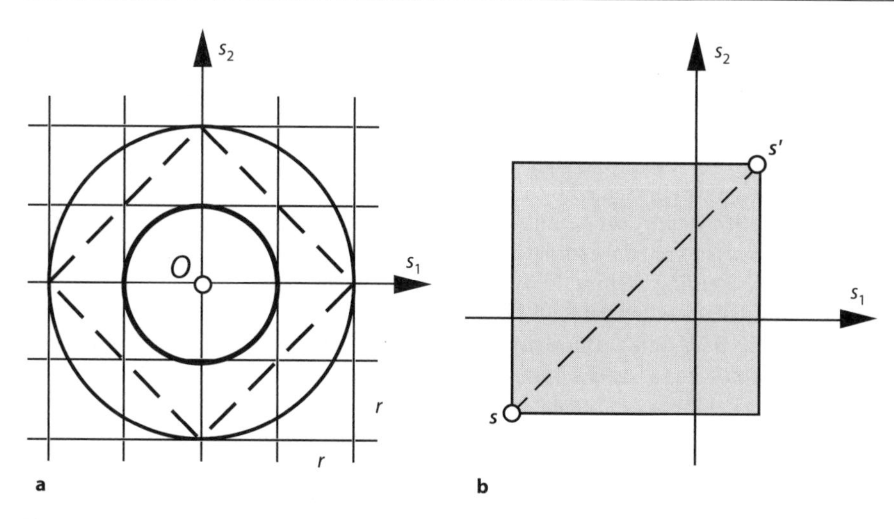

Fig. 2.8. a Iso-covariance contours for Dr. T (*solid lines*) and Dr. R. (*dashed lines*), and **b** betweeness sets for Dr. T (*dashed lines*) and Dr. R (*rectangle*)

set of points $\boldsymbol{u}$ is simply a straight line connecting $\boldsymbol{s}$ and $\boldsymbol{s}'$, whereas this set is a rectangle in the case of Dr. R (see the plots in Fig. 2.8b). ▪

Moreover, as is shown in the following example, covariance models that are permissible for Dr. T may be not so for Dr. R.

Example 2.10. The spatiotemporal covariance function (sometimes called the *Gaussian* covariance)

$$c_x(\boldsymbol{h}, \tau) = c_0 \exp(-\boldsymbol{h}^2 - v^2\tau^2) \tag{2.22}$$

where c_0 is the variance, $\boldsymbol{h} = \boldsymbol{s} - \boldsymbol{s}'$ and $\tau = t - t'$, is permissible only for Dr. T[10]. ▪

Clearly, the Euclidean metric (Eq. 2.3) is the standard metric assumed in classical geostatistics and spatial statistics applications. While in many cases this is a satisfactory assumption, there are several other applications in which the Euclidean metric is physically inappropriate and can lead to poor representations of the real phenomenon (e.g., in a fractured medium there may exist preferential flow paths along specific directions which are not represented adequately by a Euclidean metric; see also Sect. 2.1 above). As a result of the different metric structures assumed, space/time predictions obtained on the basis of the same dataset and the same mapping technique can be drastically different. And, if the Euclidean metric is not physically appropriate, the derived maps will offer a poor representation of reality. We will illustrate this substantial issue with the help of a numerical example.

[10] Mathematical proofs are given in Christakos (2000); and in Christakos and Papanicolaou (2000).

Example 2.11.[11] A realization of the (0,1)-Gaussian random field $X(s)$, $s = (s_1, s_2)$, is generated over a square area D of unit size (see Fig. 2.9a), using the widely used Cholesky decomposition simulation technique[12]. For the purposes of this study, this spatial realization is considered as the actual field. The simulated field $X(s)$ is physically characterized by a 1-norm spatial metric of the form of Eq. 2.4 with $n = 2$, and its spatial variation is represented by an exponential variogram (with a range $\varepsilon = 3$ and a sill $c_0 = 1$; both in appropriate units). In our synthetic study, Drs. T and R seek to compare two sets of spatial estimates of the field $X(s)$, which were obtained by means of the simple Kriging technique using the same dataset but different metrics[13]. In particular, the dataset available to both Drs. T and R includes 20 simulated samples (considered as hard data) arbitrarily located throughout the area D (Fig. 2.9b). As we saw above, Dr. R has a deep appreciation of the importance of properly relating a credible physical theory to the dataset available. Therefore, taking into consideration the underlying physics, Dr. R concludes (correctly) that the distances between spatial points of the field $X(s)$ within the area D are determined by the 1-norm metric. (This is consistent with the fact that the simulated field values in Fig. 2.9a exhibit a "stripping effect" along the two axes, i.e., the values tend to fluctuate more slowly along the s_1 and s_2 axes than in any other direction.) Dr. T, on the other hand, assumes (incorrectly, as it turns out) that the field $X(s)$ is characterized by a Euclidean metric, which is used by default in most commercial geostatistical packages, regardless of the physical situation (an implication of Dr. T's assumption is that there are no preferential directions in the $X(s)$ variation, etc.). As a consequence of the above considerations, in the case of Dr. R, the implementation of the simple Kriging technique relies on the 1-norm metric and leads to one set of spatial estimates (hereafter denoted by $K1$), whereas in the case of Dr. T, Kriging is based on the Euclidean metric and leads to another set of estimates (hereafter denoted by KE). Not surprisingly, the spatial variability patterns exhibited

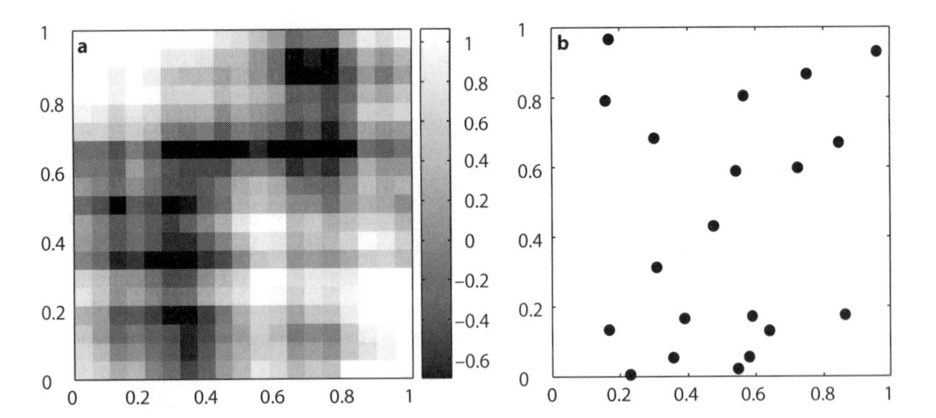

Fig. 2.9. a Simulated map; and **b** locations of the 20 hard data used in spatial estimation

[11]The reader can use the routine **example00.m** to reproduce the results of this example.
[12]For a detailed description of this random field simulation technique, see Christakos (1992).
[13]For more details about this estimation technique, see Chap. 4.

by the two sets of estimates, *K*1 and *KE*, turned out to be very different. These differences are most clearly shown in the maps of the derivatives of the *K*1 and *KE* estimates along the two axes. These maps are plotted in Fig. 2.10. In particular, the maps of the derivatives of the *K*1 estimates (Fig. 2.10c,d) show the same stripped pattern as the

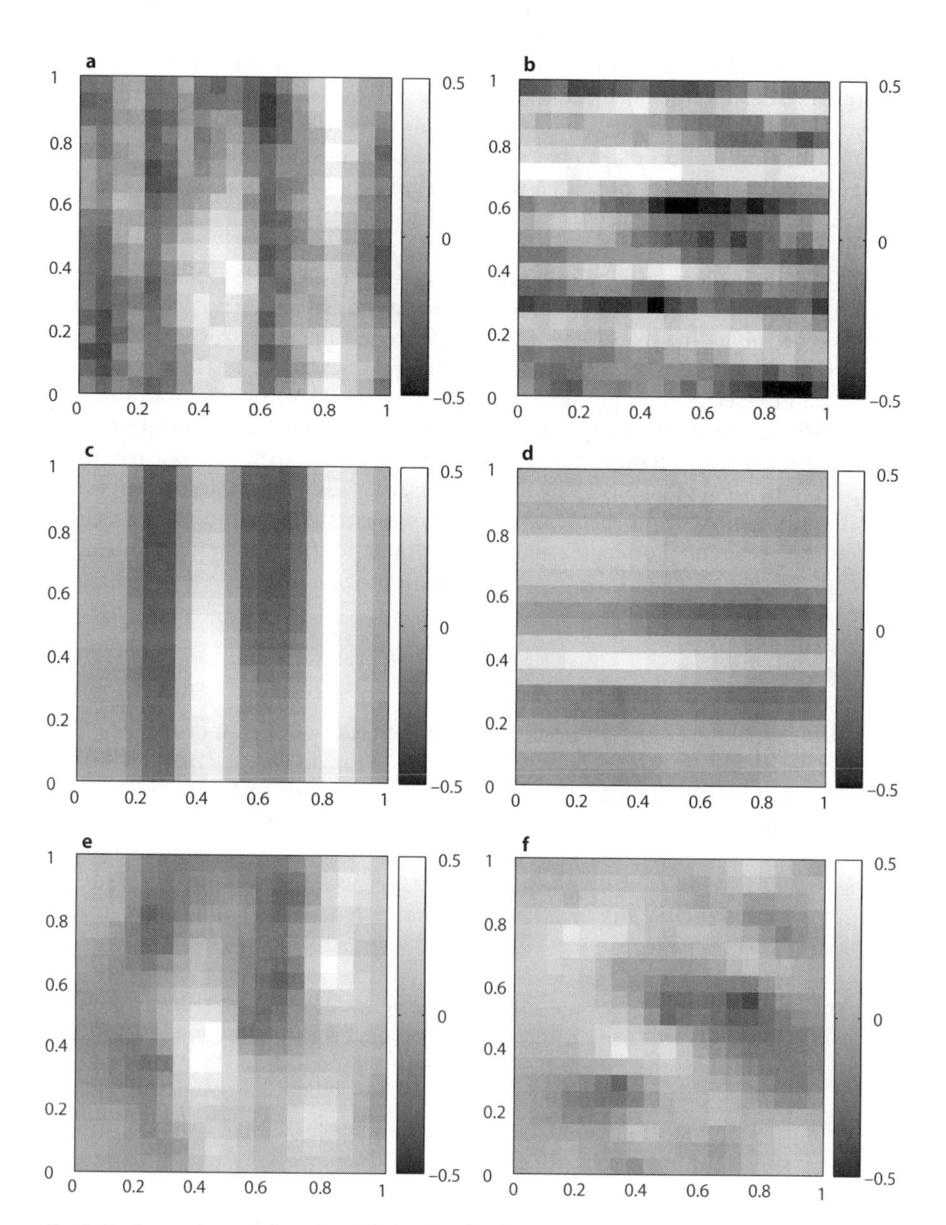

Fig. 2.10. Comparison of directional derivatives for the *K*1 and *KE* estimation maps. Left and right columns correspond, respectively, to derivatives along the vertical and horizontal axes. The 1st, 2nd and 3rd raws represent derivatives for the simulated, the *K*1, and the *KE* maps, respectively

derivatives of the simulated map (Fig. 2.10a,b), whereas the maps of the derivatives of the *KE* estimates (Fig. 2.10e,f) exhibit a different spatial variability pattern than the simulated map (e.g., variations tend to be of the same order of magnitude along any direction). The stripped pattern clearly exhibited by the derivatives of the *K*1 map is explained by the fact that – on the average – pairs of values located along the vertical axis or along the horizontal axis are expected to be statistically more similar than pairs along any other direction. The *K*1 spatial pattern is consistent with the actual one observed in the simulated maps. On the contrary, the *KE* maps fail to exhibit such a spatial pattern. In conclusion, due to the different properties of the two spatial metrics assumed, the estimation maps obtained on the basis of the same finite set of data values exhibit very different spatial patterns (the expected stripping effect for the 1-norm metric was overlooked by the *KE* approach, etc.). ▪

The rather simple numerical experiment presented in Example 2.11 above demonstrates that beyond the generation of field values at the estimation points, one should be concerned about the space/time pattern exhibited by the map. This pattern should be physically consistent with the (natural, epidemiological, etc.) field under study, whereas a misspecification of the associated metric could lead to serious problems. In the case, e.g., of conductivity variation in a porous medium, the 1-norm situation of Example 2.11 would represent a subsurface medium in which the least resistance flowpaths are oriented mainly along the two axes, whereas the Euclidean norm would correspond to an isotropic medium with no preferential flowpath directions. Overlooking the preferential directions may have a dramatic effect on subsequent studies involving these maps (e.g., scientific hypothesis testing and decision making).

2.5 Synopsis

Space/time and knowledge synthesis are two basic facets of human experience, intimately related to each other. The investigation of the main characteristics of space/time as well as their important consequences in *TGIS* analysis and modelling was the topic of the present chapter. Among other things,

Rigorous space/time modelling helps one understand the underlying physical mechanisms at work and – in the process – it improves the analytical cabapilities of geographotemporal system technology.

In the next chapter we turn our attention to the wide variety of knowledge bases available in the real world, and the ways one can obtain a meaningful synthesis of these bases for *TGIS* purposes.

Chapter 3

Knowledge Bases Synthesis

> The more facts one has, the better the judgment one can make, but one must never forget the corollary that the more facts one has, the easier it is to put them together wrong.
> *G. Heyworth*

3.1 Integrating Knowledge Bases into *TGIS*

The world around us is filled with unresolved problems and situations in which we lack a clear understanding of the underlying physical mechanisms and laws. The key to mitigating our ignorance and solving such problems lies in the physical *knowledge bases* (*KB*) we gather and their interpretation in the context of the problem at hand. In *BME* analysis, the term "knowledge base" denotes a collection of knowledge sources relevant to the problem at hand; these sources are invoked by a reasoning process aimed at solving the problem. Good *KB* should provide the right knowledge and not merely a vast range of data. Indeed, in many cases one faces the possibility of being

Swamped by data, rather than assisted by knowledge.

Because physical data are collected from several sources using different methods, conclusions are often at odds with each other, depending upon how the data are derived, organized, and presented. The organization of physical data can change its meaning and interpretation. Also, due to changes occurring across space and time, any rigorous procedure for collecting and interpreting data must take into account the contextual settings of these data. These issues are well-appreciated by the Information Technology community.

> Ideas precede our understanding of facts, although the overabundance of facts tends to obscure this. A fact can be comprehended only within the context of an idea.

says Richard Saul Wurman, in his recent book "Information Anxiety 2" (Wurman 2001).

In view of the above considerations, the *BME* approach of *Modern Spatiotemporal Geostatistics* seeks to provide a convenient framework for organizing and synthesizing one's scientific experience in terms of two distinct *KB*:

- The *general KB*, $\mathcal{G}$, which refers to background knowledge and relies on one's ability to draw on the discoveries and expertise of others (education, scientific literature, experts, etc.).
- The *specificatory* (or *case-specific*) *KB*, $\mathcal{S}$, which is obtained through experience with the specific situation (empirical observations, site-specific data, etc.).

Naturally, the *total* physical knowledge $\mathcal{K}$ available will be the union of the $\mathcal{G}$-*KB* and the $\mathcal{S}$-*KB*, i.e.,

$$\mathcal{K} = \mathcal{G} \cup \mathcal{S} \tag{3.1}$$

The choice of the appropriate *KB* to collect will depend on the specified goals of the *TGIS* study. Also, the scientific hypotheses or critical assumptions that an expert makes regarding the phenomenon in question can direct his thinking to the possible source of knowledge that is most valuable for the case. These *KB* should be organized in meaningful ways so that they are interpreted and integrated into the space/time mapping process.

The general *KB* ($\mathcal{G}$) may refer to theoretical laws and conceptual models that are usually developed for well-defined and sometimes idealized conceptual environments. This *KB* is instrumental in understanding the underlying physical mechanisms, discovering unity behind apparently dissimilar phenomena, and putting aside complexities that are beside the point in question[1]. On the other hand, when we move beyond theoretical laws in controlled environments, reality is frequently complicated by site-specific *KB* ($\mathcal{S}$), such as macro geological effects creating discontinuities in soil parameter variations. Dispersal and retention of subsurface contamination, e.g., may vary significantly from one soil type to another. Also, physical properties (like porosity, hardness, and void ratio) change abruptly between adjacent geological units. In light of these considerations, Fig. 3.1 presents several possible ways to classify the various knowledge synthesis components that may be considered in a *TGIS*. Among the various classifications, special emphasis is given to the following:

- Multi-disciplinary *KB* (i.e., involving physical, medical, biological, epidemiological, and other scientific disciplines).
- *KB* based on different instruments of observation and experimentation.
- *KB* represented in different space/time geometries.
- *KB* originated in different physical and population scales.

In a world dominated by interconnected information technology systems, *TGIS* specialists need to offer their wisdom in small, manageable chunks. However, simply to digitize thoughts and offer them to *TGIS* users as large arrays of numbers risks ab-

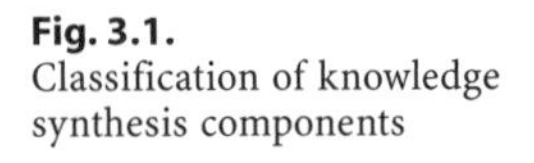
Fig. 3.1.
Classification of knowledge synthesis components

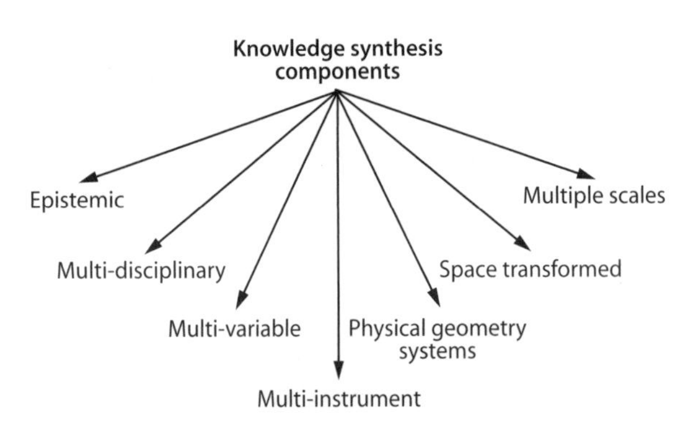

[1] To refer to a historical example, when Galileo was trying to understand the fundamentals of objects in motion, he knew that the key issue was to conceptualize how objects fall through a vacuum, ignoring wind effects and other details. If he could get the basic equations right for this conceptual environment, he could later account for other details.

stracting them from their multilayered conceptual and epistemic contexts. Therefore, most *TGIS* specialists are well aware of the paramount significance of the following dictum:

An adequate scientific study must be based on a physically meaningful synthesis *of general and site-specific KB.*

This dictum is valid in a wide range of scientific disciplines. Recenly, Steinberg and Kareiva (1997) have quite appropriately emphasized the importance of integrating general theories with site-specific empirical findings:

The greatest challenge facing 'spatial ecology' is wedding the pertinent theory to empirical research.

In their analysis of Serrengeti wildbeest, Hilborn and Mangel (1997) combined different forms of general knowledge (logistic and life history models, etc.) with various sources of diverse data (abundance estimates, etc.) in order to address issues such as the level of illegal harvest and the potential response of the herd to harvest increases. In subsurface contamination studies (e.g., Spitz and Moreno 1996), the region is properly divided into control cells, each with its own hydrogeologic characteristics (boundary and initial conditions, faults, dykes, sources, sinks, topographic contours, etc.), and physical laws are applied to each cell with the help of numerical models and computers, generating predictions of contaminant distributions. From the viewpoint of scientific methodology, this approach combines reductionistic with holistic elements: first, the modeller studies parts of the natural system in isolation from the rest, and then graduates successfully to the holistic complexities of the whole[2]. In the subsequent sections we will discuss the structure of the *KB*, $\mathcal{G}$ and $\mathcal{S}$, in more detail.

3.2 General *KB* and the Associated Physical Constraints

As we observed in the previous chapter, knowledge in dynamic interplay with space/time can offer valuable insight into scientific inquiry. In many *TGIS* applications, the general $\mathcal{G}$-*KB* is conveniently expressed in terms of a set of *stochastic equations* expressing knowledge in space and time, as follows:

$$\begin{aligned}\overline{h_\alpha}(\boldsymbol{p}_{\text{map}}) &= \overline{g_\alpha(\boldsymbol{x}_{\text{map}})} \quad \text{with} \\ \overline{g_\alpha(\boldsymbol{x}_{\text{map}})} &= \int d\boldsymbol{\chi}_{\text{map}}\, g_\alpha(\boldsymbol{\chi}_{\text{map}}) f_{\mathcal{G}}(\boldsymbol{\chi}_{\text{map}})\end{aligned} \tag{3.2}$$

where g_α and h_α ($\alpha = 0,1,\ldots,N$) are sets of known functions of $\boldsymbol{\chi}_{\text{map}}$ and $\boldsymbol{p}_{\text{map}}$, and N is the number of moment equations considered. The right-hand side of Eq. 3.2 depends

[2] Drawing an analogy is illustrating. When engineers send a rocket ship to Space, they do not expect to reach a target with just one implementation of the theoretical laws of mechanics and a unique set of initial conditions. Instead, with the help of computers, they rely on the continuous incorporation of the changing conditions of the ship's motion (velocity changes due to gravitational fields from other objects, etc.) into their calculations of the theoretical laws and the path is periodically corrected by the application of rocket engines.

on the χ_{map}-values, the $\boldsymbol{p}_{\text{map}}$-coordinates, and the pdf f_G associated with the general knowledge $\mathcal{G}$. The left-hand side represents stochastic expectations of the (natural, epidemiological, etc.) fields involved. By convention, $g_0 = 1$, so that $\overline{g_0} = 1$ is a normalization constant. The g_α's ($\alpha > 0$) are chosen so that the stochastic expectations $\overline{h_\alpha}$ can be calculated from field data or inferred from physical laws, empirical charts, etc. Thus,

Eqs. 3.2 may be viewed as physical constraints which account for a variety of general KB

including

- Statistical correlation functions (means, covariances, variograms, multiple-point moments, non-linear statistics etc.).
- Scientific models (physical laws, phenomenological relationships, biological theories, etc.).

The g_α and h_α functions do not necessarily have the same mathematical form (see Examples 3.5 and 3.6 below). Various other general *KB*s can be found in the *BME* literature. In this book, however, we chose to focus on the two *KB* above.

3.2.1 Space/Time Correlation Functions Between Two or More Points (Multiple-Point Statistics)

In many situations in *TGIS* practice, a salient issue concerning the integration of space/time data is to establish a suitable format to correlate this data. By imposing a logical format on a set of raw data, space/time correlation functions of various forms, such as

- *Covariance*
- *Variogram*
- *Trivariance*, and
- Higher order *multiple-point* statistics (linear and non-linear)

are able to generate valuable information about the spatiotemporal variation of the physical data. An extensive list of correlation functions (space/time separable and non-separable) currently in use can be found in the *BME* literature. Among the more rewarding pay-offs of *BME*-based advanced *TGIS* functions is that,

The BME approach provides TGIS with the means to incorporate a multiple-point space/time statistics of any order, linear or non-linear, in a straightforward yet rigorous manner,

which is not the case with most previous mapping techniques, like classical Kriging and statistical regression. All that is needed, in the case of *BME*, is to formulate the g_α-functions corresponding to the multiple-point statistics of interest, and then follow the basic steps of the mapping approach: *BME* will automatically incorporate multiple-point statistics of any order (see also *Steps* 1 through 6 in Sect. 4.2.1 of Chap. 4).

For most *TGIS*, the correlation functions characterizing the spatiotemporal variation of the phenomenon in question constitute an important general *KB*, $\mathcal{G}$. The formulation of the g_α-functions associated with the correlation functions of $\mathcal{G}$ is best illustrated with the help of a few examples[3].

Example 3.1. If we choose $g_\alpha(\chi_i) = \chi_i^\lambda$ (λ is a positive integer), Eq. 3.2 gives the *λ-th order moment*

$$\overline{g_\alpha(x_i)} = \overline{x_i^\lambda} \tag{3.3}$$

of the random field at each point $\boldsymbol{p}_i$. By choosing $g_\alpha(\chi_i,\chi_j) = ½(\chi_i - \chi_j)^2$ for various pairs of points, Eq. 3.2 gives the *variogram* function,

$$\overline{g_\alpha(x_i,x_j)} = \frac{1}{2}\overline{(x_i - x_j)^2} = \gamma_x(\boldsymbol{p}_i,\boldsymbol{p}_j) \tag{3.4}$$

between any two points $\boldsymbol{p}_i$ and $\boldsymbol{p}_j$ in the space/time domain. Figure 3.2 gives a schematic illustration of the calculation of the variogram values for the *one-dimensional* random field $X(s)$ at spatial lags $h_1 = s_{i+1} - s_i$ and $h_{10} = s_{i+10} - s_i$ (Fig. 3.3a). The original realization of the random field is denoted as χ_i, whereas the lagged realizations are denoted as χ_{i+m} (m = 1 and 10). The estimated $\gamma_x(h_m)$ value of the one-dimensional variogram is equal to one-half the arithmetic average of the squares of all local $|\chi_{i+m} - \chi_i|$-values (the magnitudes of the latter are schematically represented by the vertical lines). As the representation of Fig. 3.2 shows, within certain ranges the variogram $\gamma_x(h_m)$ is an increasing function of the lag h_m (see also Fig. 3.3a), although there are situations in which it may also involve a periodic component.

In a separate case, a contour-based representation of a typical variogram in two spatial dimensions, (s_1, s_2), is given in Fig. 3.3b. The contour values indicate the

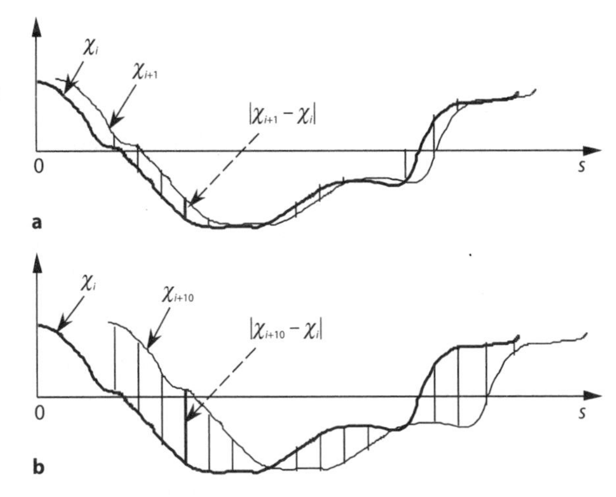

Fig. 3.2. Schematic calculation of the variogram values for the one-dimensional random field at lag: **a** $h_1 = s_{i+1} - s_i$; and **b** $h_{10} = s_{i+10} - s_i$

[3] Although, due to lack of space, not a detailed exposition is given in this book of the several computational issues related to the estimation of these functions in practice. Naturally, these issues have been taken into consideration in the development of the *BMElib* computer software (Chap. 7). For additional details the interested reader is referred to the existing geostatistical literature.

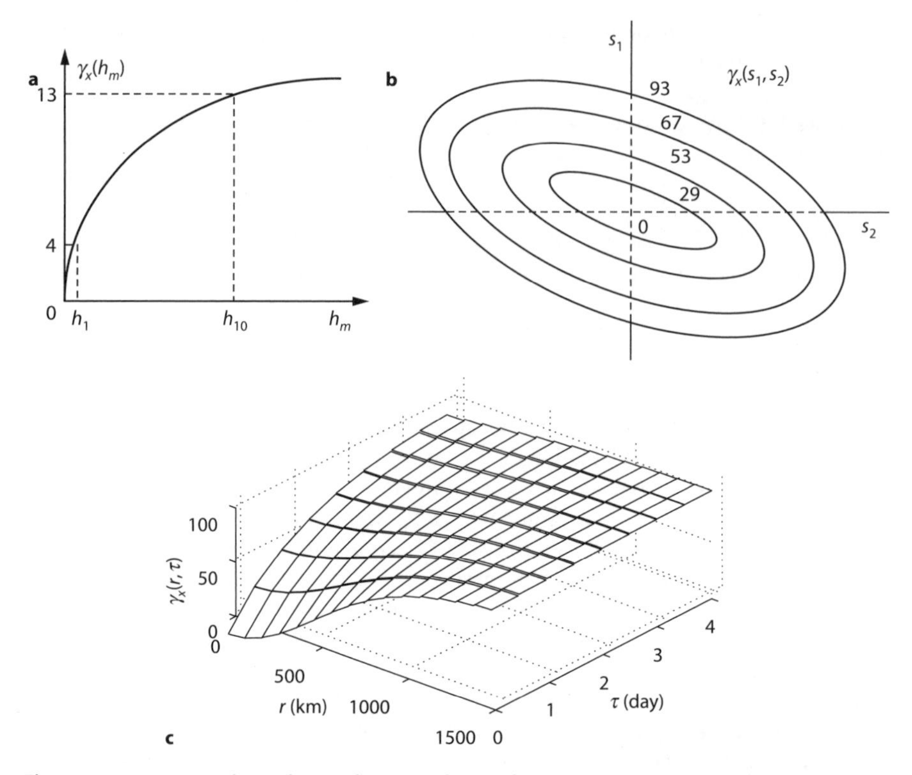

Fig. 3.3. **a** A variogram plot in the one-dimensional space; **b** a contour representation of a variogram in the two-dimensional space; **c** ozone concentration variogram (in Dobson^2) in the space/time domain

variogram values between a point located at the origin o and other points in the two-dimensional domain. Finally, a spatiotemporal variogram $\gamma_x(r, \tau)$ (in Dobson^2) of ozone concentration is displayed in Fig. 3.3c, where r and τ are, respectively, the spatial and temporal distances between two points in the space/time domain. This variogram was calculated from TOMS (Total Ozone Mapping Spectrometer) data; the spatial and temporal correlation ranges are, respectively, $a_r = 12$ km and $a_\tau = 4$ days. ▪

The covariance and variogram functions above are both two-point statistics. In certain cases (nonlinear systems, complex natural variations, etc.), spatiotemporal analysis and mapping may be improved by incorporating multiple-point statistics (i.,e., correlation functions involving three or more points in space/time). The following example shows how straightforward it is to formulate the g_α-functions associated with multiple-point statistics, for subsequent use in the *BME* space/time mapping approach. Some interesting properties of the trivariance function are also explored.

Example 3.2. By letting $g_\alpha(\chi_i, \chi_j, \chi_k) = \chi_i \chi_j \chi_k$, Eq. 3.2 defines the (non-centered) *trivariance* function (*3-point* space/time statistic),

$$\overline{g_\alpha(x_i, x_j, x_k)} = \overline{x_i x_j x_k} \tag{3.5}$$

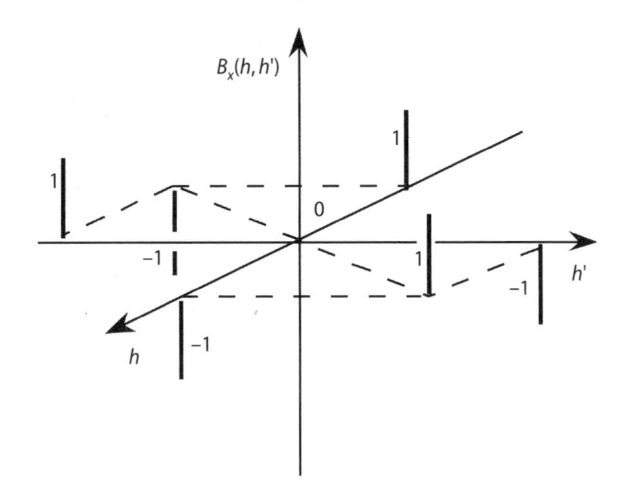

Fig. 3.4. An illustration of the spatial trivariance function (3-point statistic)

between three points, $\boldsymbol{p}_i$, $\boldsymbol{p}_j$ and $\boldsymbol{p}_k$. Next, a simple numerical illustration of the trivariance function is obtained as follows. Let $x_s = u_s - u_{s-1}$, where u_s is a zero mean field with $\overline{u_s^2} = \overline{u_s^3} = 1$ and $\overline{u_s u_{s+h} u_{s+h'}} = \delta_{h,h'}$, where $\delta_{h,h'}$ is the two-dimensional Kronecker delta function (Nikias and Petropulu 1993). In this case, the trivariance function is given by $B_x(h, h') = \overline{x_s x_{s+h} x_{s+h'}} = \delta_{h-1,h'-1} + \delta_{h+1,h'} - \delta_{h+1,h'+1} - \delta_{h-1,h'} - \delta_{h,h'-1} + \delta_{h,h'+1}$. This simple trivariance is plotted in Fig. 3.4. It is worth noticing that at the origin, $B_x(0, 0) = 0$. Moreover, one should also point out some interesting symmetry properties of the trivariance function, i.e., it is valid that, $B_x(h, h') = B_x(h', h) = B_x(-h', h - h') = B_x(h' - h, -h) = B_x(h - h', -h') = B_x(-h, h' - h)$.

In a separate case, we let $g_\alpha(\chi_i, \chi_j, \chi_k, \chi_l) = \chi_i \chi_j^3 \chi_k^5 \chi_l^2$. Then, Eq. 3.2 gives the *four-point non-linear statistic*

$$\overline{g_\alpha(x_i, x_j, x_k, x_l)} = \overline{x_i x_j^3 x_k^5 x_l^2} \tag{3.6}$$

for the points $\boldsymbol{p}_i$, $\boldsymbol{p}_j$, $\boldsymbol{p}_k$ and $\boldsymbol{p}_l$. Higher order statistics can be defined in a similar fashion. In this example, the g_α and h_α functions have the same mathematical form (which is not, though, generally the case; see Example 3.5 below). ▪

As it turns out, certain multiple-point statistics can have significant differences from two-point statistics. E.g., it is possible that a field is homogeneous/stationary in its two-point statistic but non-homogeneous/non-stationary in its three-point statistic (trivariance).

Example 3.3. Generally, a random field is considered as spatially homogeneous with respect to a multiple-order statistic if the latter is only a function of the vector distances between any two points. Thus, if $X(\boldsymbol{s})$ is homogeneous in the (non-centered) *tetravariance* (*four-point* statistic) the latter is expressed as

$$\Theta_x(\boldsymbol{h}, \boldsymbol{h}', \boldsymbol{h}'') = \overline{X(\boldsymbol{s})X(\boldsymbol{s}+\boldsymbol{h})X(\boldsymbol{s}+\boldsymbol{h}')X(\boldsymbol{s}+\boldsymbol{h}'')} \tag{3.7}$$

Let $\theta_x(\boldsymbol{h},\boldsymbol{h}',\boldsymbol{h}'') = \overline{[X(\boldsymbol{s}) - m_x][X(\boldsymbol{s}+\boldsymbol{h}) - m_x][X(\boldsymbol{s}+\boldsymbol{h}') - m_x][X(\boldsymbol{s}+\boldsymbol{h}'') - m_x]}$ be the centered tetravariance, where m_x is the space-independent field mean. Then, it is easily shown that

$$\begin{aligned}\theta_x(\boldsymbol{h},\boldsymbol{h}',\boldsymbol{h}'') = {} & \Theta_x(\boldsymbol{h},\boldsymbol{h}',\boldsymbol{h}'') - m_x[B_x(\boldsymbol{h},\boldsymbol{h}') + B_x(\boldsymbol{h},\boldsymbol{h}'') + B_x(\boldsymbol{h}',\boldsymbol{h}'') + B_x(\boldsymbol{h}'-\boldsymbol{h},\boldsymbol{h}''-\boldsymbol{h})] \\ & + m_x^2[C_x(\boldsymbol{h}) + C_x(\boldsymbol{h}') + C_x(\boldsymbol{h}'') + C_x(\boldsymbol{h}'-\boldsymbol{h}) + C_x(\boldsymbol{h}''-\boldsymbol{h}) + C_x(\boldsymbol{h}''-\boldsymbol{h}')] \\ & - 3m_x^4 \end{aligned} \tag{3.8}$$

On the other hand, $c_x(\boldsymbol{h}) = C_x(\boldsymbol{h})$ and $\beta_x(\boldsymbol{h},\boldsymbol{h}') = B_x(\boldsymbol{h},\boldsymbol{h}')$, where $c_x(\beta_x)$ and $C_x(B_x)$ denote the non-centerd and centerd covariance (trivariance) of $X(\boldsymbol{s})$, respectively. Furthermore, let us consider the field $x_s = u_s\cos(as)$, where u_s is zero mean homogeneous field independent to x_s such that $\overline{u_s u_{s+h}} = C_u(h)$ and $\overline{u_s u_{s+h} u_{s+h'}} = B_u(h,h')$, and a is a constant. Then, the covariance of x_s is $\overline{x_s x_{s+h'}} = C_u(h)\cos(ah) = C_x(h)$, which is a function of h only. On the other hand, the trivariance is $\overline{x_s x_{s+h} x_{s+h'}} = B_u(h,h')\cos(ah)\cos(as+ah)\cos(as+ah') = B_x(s,h,h')$, which is a function of h, h' and s. Hence, the field x_s is homogeneous in the covariance but not in the trivariance. ▪

The above example shows that new multiple-point covariance models can be generated from a set of known models using the (empirical or theoretical equations) relating the corresponding fields. In geostatistical applications involving indicator random fields, multiple-point statistics were suggested by Christakos and Hristopulos (1996, 1998).

Example 3.4. In environmental applications, it is sometimes useful to calculate multiple-point statistics of indicator random fields $I_x(\boldsymbol{s},z)$: $I_x(\boldsymbol{s},z) = 1$, if $X(\boldsymbol{s}) \geq z$, $= 0$, otherwise. The three-point spatial indicator is defined as $B_I(\boldsymbol{s},\boldsymbol{s}',\boldsymbol{s}'',z) = \overline{I_x(\boldsymbol{s},z)I_x(\boldsymbol{s}+\boldsymbol{h},z)I_x(\boldsymbol{s}+\boldsymbol{h}',z)}$. In the case of spatial homogeneity and if z is equal to the median of $X(\boldsymbol{s})$, the three-point statistics can be expressed as

$$B_I(\boldsymbol{h},\boldsymbol{h}';z) = \tfrac{1}{2}[1 - \gamma_I(\boldsymbol{h};z) - \gamma_I(\boldsymbol{h}';z) - \gamma_I(\boldsymbol{h}-\boldsymbol{h}';z)] \tag{3.9}$$

where γ_I is the indicator variogram. Higher order multiple-point indicator statistics can be defined in a similar fashion. ▪

One point that needs special emphasis is that the above examples reveal the substantial ability of the *BME* formalism to

Account for multiple-point statistics (linear and non-linear) in a rigorous and efficient manner.

This is a crucial point, since in certain *TGIS* applications the one-point and two-point statistics (like the mean, the covariance, or the variogram) may offer a rather limited representation of the actual space/time variation.

Comment 3.1. It is worth mentioning that the use of raw data to obtain useful information in the form of space/time correlation functions has an epistemic priority at this stage. At a later stage, the data may be used with another objective in mind (e.g., space/time estimation; see Chap. 4).

3.2.2 Physical Models

Scientists often stress the importance of specifying a model of the relation of data to the hypotheses under investigation. A model is expressed in terms of equations which correspond to real physical, biological, etc. situations. Without such a model, the data cannot be interpreted.[4] A familiar slogan from the practice of life support sciences makes this clear:

No physical model, no scientific inference.

This slogan should not be forgotten when one investigates *TGIS* methods aiming at the visual representation of this practice. In fact, most natural phenomena governed by field equations include situations that need to be treated from a *stochastic* viewpoint. More specifically, in situations involving uncertain elements and random fluctuations, one formally casts the governing physical equations into a stochastic form that may involve random field realizations, probability distributions, or space/time moments. As a result of their physical basis, these stochastic equations provide the means for sound scientific inferences, as opposed to merely statistical inferences (in terms of minimum variance, bias, efficiency, estimation and confidence tests, etc.; Bury 1975).[5]

Mathematical models of physical laws make natural partners for *TGIS*, given the latter's capabilities in terms of data integration and visualization. For instance, *TGIS* make possible the integration of the different data sources considered in environmental equity analysis, including data about the locations of hazardous sites (toxic release inventory sites, Superfund sites, etc.) and data on the characteristics (race, age, social status, etc.) of the exposed population. The geographical plume analysis of Maitin and Klaber (1993) is based on the integration of *GIS* with chemical dispersion models (which account for information about weather conditions, pollutant type and amount, etc.). This integration allows decision-makers to obtain useful scientific estimates of the impact of a released toxic substance on a given population. Spatial decision support systems merge *GIS* technology with mathematical models of environmental burden, thus providing the decision-maker with the means to analyze a hazardous situation relative to certain optimal objectives (cost minimization, optimizing social consequences, etc.) and pollution levels (Baiamonte 1996). Heuvelink (1998) gives a review of *GIS* studies involving physically based models of distributed dynamic catchments, soil erosion processes, and water resources management. Moreover, Malczewski (1999) discusses several real-world decision situations which give rise to *GIS*-based multicriteria decision making.

[4] In a metaphorical sense, one reads physical evidence as an insider, not as a visiting Martian would read it.

[5] Celebrated early stochastic modelling approaches based on physical laws include Maxwell's and Boltzmann's development of the kinetic theory of gases (in 1860 and 1896), Planck's derivation of the radiation law (in 1900), Gibbs' formulation of statistical mechanics (in 1901), Einstein's and Langevin's analyses of Brownian motion (in 1905 and 1908), Taylor's and von Karman's theories of turbulent motion (in 1921 and 1937), and Heisenberg's and Born's approaches to modern quantum mechanics (in 1925 and 1926).

Modern *TGIS* analysis attributes great importance to the incorporation of natural laws into the space/time mapping process. One of *BME*'s main attractions, then, is

BME's ability to incorporate natural laws into the TGIS mapping process in a rigorous and efficient manner.

Therefore, the use of powerful scientific laws (physical, ecological, epidemiological, etc.) can offer important geographotemporal insights that could not have been obtained by means of descriptive approaches aiming merely to portray observed patterns. The reader is referred to Sect. 6.3.4, in which we discuss a numerical example which deals with the successful implementation of computational *BME* analysis in the study of the advection-reaction equation governing the distribution of pollutant concentration along a river at various time instants. Furthermore, two analytical examples are briefly presented below, which illustrate the *BME*-based formulation of scientific laws so that they can be properly incorporated into *TGIS* modelling and mapping[6]. The readers unfamiliar with stochastic partial differential equations may prefer to skip these two examples.

Example 3.5. In hydrogeologic sciences, the distribution of the soil moisture field, $X(\boldsymbol{p})$, is assumed to obey the following physical law (Entekhabi and Rodriguez-Iturbe 1994)

$$\eta Z_r \frac{\partial}{\partial t} X(\boldsymbol{p}) = -\eta X(\boldsymbol{p}) + \kappa \nabla^2 X(\boldsymbol{p}) + Y(\boldsymbol{p}) \tag{3.10}$$

where η is the soil porosity, Z_r is the depth of active soil, κ is a coefficient that is space-independent, and $Y(\boldsymbol{p})$ represents the rainfall process in space/time with known statistics (mean m_y, covariance c_y, etc.). The 1st- and 2nd-order stochastic moment equations associated with the law 3.10 can be expressed in terms of the g and h functions as follows,

$$\begin{aligned}
\overline{h_1} &= \eta Z_r \frac{\partial}{\partial t} \overline{X(\boldsymbol{p})} \\
\overline{g_1} &= \int d\chi (-\eta\chi + \kappa\chi\nabla_s^2) f_{\mathcal{G}}(\chi) + m_y \\
\overline{h_2} &= \eta Z_r \frac{\partial}{\partial t'} C_x(\boldsymbol{p}, \boldsymbol{p}') \\
\overline{g_2} &= \iint d\chi d\chi' \chi\chi' (-\eta + \kappa\nabla_{s'}^2) f_{\mathcal{G}}(\chi, \chi') + \iint d\chi d\psi' \chi\psi' f_{\mathcal{G}}(\chi, \psi')
\end{aligned} \tag{3.11}$$

for all points $\boldsymbol{p}$ and $\boldsymbol{p}'$ of interest, where $C_x(\boldsymbol{p}, \boldsymbol{p}')$ is the space/time (non-centered) covariance of the soil moisture field, and $f_{\mathcal{G}}$ are the corresponding general knowledge-based pdf. In fact, space/time variograms and higher order moments of the soil moisture field (such as, 3rd, 4th or any other order single point moment), as well as multiple-point statistics resulting from the physical law 3.10, can be expressed in a similar manner. ■

[6] A detailed treatment of the subject of physical law-based *BME* analysis can be found in Christakos (2000).

Example 3.6. According to the spatiotemporal Kermack-McKendrick law of communicable disease (see e.g., Holmes 1997), susceptible individuals become infected by contact with surrounding infected individuals according to the following set of differential equations,

$$\begin{aligned} \tfrac{\partial}{\partial t} X(\boldsymbol{p}) &= -\beta\eta(\boldsymbol{p})X(\boldsymbol{p}) \\ \tfrac{\partial}{\partial t} Y(\boldsymbol{p}) &= \beta\eta(\boldsymbol{p})X(\boldsymbol{p}) - \lambda Y(\boldsymbol{p}) \\ \tfrac{\partial}{\partial t} Z(\boldsymbol{p}) &= \lambda Y(\boldsymbol{p}) \end{aligned} \tag{3.12}$$

where the random fields $X(\boldsymbol{p})$, $Y(\boldsymbol{p})$, and $Z(\boldsymbol{p})$ denote the proportions of susceptible, infected, and resistant (i.e., immune) individuals, respectively; $\eta(\boldsymbol{p})$ is a weighted function of the number of infected individuals within a contact radius r of a susceptible individual; and λ denotes the rate at which individuals recover and become immune, whereas β expresses the rate at which susceptible individuals become infected. The mean value equations associated with the epidemiologic laws 3.12 can be expressed in terms of the g and h functions as follows,

$$\begin{aligned} \overline{h_1} &= \tfrac{\partial}{\partial t}\overline{X(\boldsymbol{p})}, \quad \overline{h_2} = \tfrac{\partial}{\partial t}\overline{Y(\boldsymbol{p})}, \quad \overline{h_3} = \tfrac{\partial}{\partial t}\overline{Z(\boldsymbol{p})} \\ \overline{g_1} &= -\beta\eta \int d\chi\chi f_{\mathcal{G}}(\chi), \quad \overline{g_2} = \beta\eta \int d\chi\chi f_{\mathcal{G}}(\chi) - \lambda \int d\psi\psi f_{\mathcal{G}}(\psi) \\ \overline{g_3} &= \lambda \int d\psi\psi f_{\mathcal{G}}(\psi) \end{aligned} \tag{3.13}$$

for all space/time points $\boldsymbol{p}$ of interest. As in the previous example, higher order statistical moments in space and time are formulated in a similar fashion. ▪

As the previous examples demonstrate, another valuable capability of the *BME* approach is that it allows the *TGIS* expert to

Obtain a good idea about the functional form of the space/time statistics (covariance, variogram, trivariance, etc.) using important information sources, like the powerful laws of life support sciences.

This *BME* feature can be extremely useful since, as is well-known among geostatisticians, the construction of space/time statistics from sparse data is a difficult affair that often leads to serious inference problems.

Comment 3.2. In light of the above considerations, a distinction should be made between the *physical* model-based mapping discussed here, and the *statistical* model-based mapping (see e.g., Diggle et al. 1998). The former refers to natural laws and is concerned with scientific inference, whereas the latter refers to statistical models and focuses on statistical inference.

The model-based knowledge of the *BME* formulation above can be integrated with other kinds of case-specific knowledge to generate pdf of the predicted field values at any desired point across space and time. As we shall see in the numerical application of Sect. 6.3.4, the general knowledge provided by the advection-reaction equation (which models the temporal pollutant concentration distribution along a river) is combined with case-specific probabilistic evidence to yield the pollutant concentration pdf along the river.

3.3 Specificatory *KB*

Specificatory or *case-specific S-KB* is knowledge about the specific situation. Unlike the general *KB*, the case-specific *S* refers to a particular occurrence of state of affairs at a particular location and at a particular time[7]. The knowledge of a particular system depends considerably on the way in which the system is observed. In this respect, the assignment of case-specific *observables* (such as natural fields, topographic parameters, economic and social variables) is a vital part of the study of a system. From a *TGIS* point of view, site-specific information about these observables is often available in the form of

a *Base maps.*
b *Attribute data bases.*

A base map is a graphic representation of the geographic layout (e.g., the topography of a region, rivers, streets, and census-tract boundaries). Attribute datasets describe natural or epidemiological fields (subsurface contaminant values, facies categories, petrophysical properties, seismic data, air pollutant patterns, breast cancer incidence distribution, demographics of a region, etc.). Base maps may be used in conjunction with attribute datasets. A host of data sources are employed by *TGIS* to obtain *KB* of the form (*a*) and (*b*) above. Primary sources include field data, tabular data, reports, and any kind of map produced by various organizations. Advanced data sources include *remote sensing* (i.e., a collection of landscape and other data by low- and high-altitude aircrafts, as well as satellites and space-shuttles).

Example 3.7. Base maps showing all streets and census-tract boundaries for the U.S. are provided by government-produced topographically integrated geographic encoding and referencing files (ESRI 1990). Attribute datasets are available through government agencies like the U.S. Geological Survey, Environmental Protection Agency, Bureau of Census, and the internet, state offices, and professional organizations. Also, attribute datasets are obtained from local surveys or sampling networks developed for the specific problem. ▪

Reality is frequently complicated by large scale *site-specific characteristics*, such as macro geological effects (faulting, erosion, folding, etc.), which have created considerable discontinuities in the variations of the regionalized physical fields of interest. This situation is illustrated with the help of the following example.

Example 3.8. A common case of geological faulting and erosional complexity is displayed in Fig. 3.5. Typically, the most significant complications are created by cycles of tectonic activity that have caused buckling, folding, shearing, faulting, etc. The dispersal and retention of subsurface contamination may vary significantly from one soil type to another. Physical properties, such as porosity, hardness, void ratio, and fracture density, could change abruptly between adjacent geological units or strata. ▪

[7] In the view of many geographers, the foundations of modern geographical information systems were laid down during the period of the Enlightenment, when careful case-specific observations and measurements achieved high status in many scientific disciplines (Robinson et al. 1995).

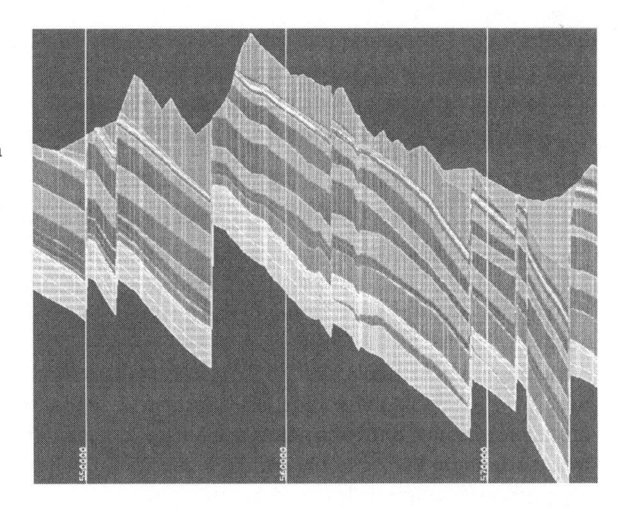

Fig. 3.5.
A vertical East-West section through a volume data structure representation of subsurface structure and stratigraphy for the proposed Yucca mountain high-level waste facility in Nevada (from Houlding 2000)

The synthesis of diverse knowledge sources such as the above, which may occur at several space/time physical scales, can be a difficult yet very important task of modern *TGIS* modelling and mapping. At the same time,

It is the unique capability of TGIS to complete the synthesis task successfully that makes it an indispensable technology in a wide variety of real-world problems.

3.3.1 Hard and Soft Data

From the point of view of scientific methodology (see also Sect. 8.1), a natural system is generally described by

i. Specifying the appropriate *observables* for the situation.
ii. Characterizing the *linkages* between these observables.

Aspect (*i*) helps the *TGIS* specialist make the necessary contact with reality. Aspect (*ii*) provides the means for building meaningful physical models, estimating a difficult or expensive-to-measure observable in terms of another observable which is easy and inexpensive to measure, etc. In many cases, the inaccurate predictions obtained for a natural system are due either to poorly measured observables or to wrongly chosen observables for that system.

The choice of the right observables for the situation depends, to a large extent, on the proper conceptualization of the life-support system and the use of the adequate physical models (see discussion in Sect. 3.2 above). On the other hand, there exist various procedures employed in gathering physical data for the observables under consideration. In fact, one could establish a *continuum* which starts with direct sensory observations and proceeds to significantly complex, indirect observation procedures. The geologist gathers data by way of field experiments, a biologist by using the microscope, an epidemiologist by conducting interviews with subjects in a population, etc. In surveying (Wolf and Brinker 1989), measurements may be made directly (e.g., by applying a tape to line, fitting a protractor to an angle, or turning an angle with a transit or theodolite), or indirectly with the help of relationships between the attribute of interest and some other known values (e.g., indirect measurements are made necessary when it is not possible to apply a measuring instrument directly to the distance or the angle to be measured).

Early in the design of spatiotemporal analysis and mapping the modeller must make decisions regarding the availability and accessibility of the specificatory *KB* required. The users of *BME* techniques are, in fact, well aware of an important concern, namely, that they should

Always plan the data gathering procedure in a purposeful way,

i.e., one should seek to yield data relevant to the specific problem under consideration. So, depending on the situation at hand, different kinds of data and experimental designs will be more or less appropriate.

In view of the above considerations, the selection of an adequate case-specific dataset for these observables is a vital part of knowledge acquisition. For the purpose of *TGIS* analysis, the attribute datasets $\boldsymbol{\chi}_{\text{data}}$ considered as specificatory *KB* are usually divided into two main categories, as follows

$$\mathcal{S}\colon \boldsymbol{\chi}_{\text{data}} = (\boldsymbol{\chi}_{\text{hard}}, \boldsymbol{\chi}_{\text{soft}}) = (\chi_1, \ldots, \chi_m) \tag{3.14}$$

where

- $\boldsymbol{\chi}_{\text{hard}}$ denotes *hard* data (accurate measurements obtained from real-time observation devices, numerical simulation, etc.), and
- $\boldsymbol{\chi}_{\text{soft}}$ denotes *soft* data (uncertain observations expressed in terms of interval values, probability statements, empirical charts, etc.).

The hard data vector $\boldsymbol{\chi}_{\text{hard}}$ represents measurements obtained with the help of instruments which, for all practical purposes, are considered accurate (i.e., they produce negligible measurement errors). We assume that the hard data available at a set of m_h ($<m$) points are represented as follows:

$$\mathcal{S}\colon \boldsymbol{\chi}_{\text{hard}} = (\chi_1, \ldots, \chi_{m_h}) \tag{3.15}$$

The *KB* 3.15 includes single-valued measurements χ_i ($i = 1, \ldots, m_h$) in space/time. These values could be the results of experimentation. As the Ancients used to say,

La experimencia madre es de la ciencia.

(I.e., "the experiment is the mother of science".) Indeed, because of *TGIS*' reliance on experimental evidence, great value is placed on the development of better instruments and techniques of observation. As already mentioned (see Comment 3.1), sometimes the set of hard data may be used with different epistemic objectives in mind. Also, *TGIS* modelling will not be effective if these data are not informative and the problem formulation is not physically meaningful.

TGIS modelling and mapping also include incomplete and/or qualitative observation statements linked to experts' opinions, experience, intuition, questionnaires, equipment shortcomings, etc. These observation statements take the form of a soft data vector $\boldsymbol{\chi}_{\text{soft}}$ which will be assumed available at the remaining $m_s = m - m_h$ points, i.e.,

$$\mathcal{S}: \boldsymbol{\chi}_{\text{soft}} = (\chi_{m_h+1}, \ldots, \chi_m) \tag{3.16}$$

TGIS practitioners continuously seek ways to extract the requisite soft data vectors $\boldsymbol{\chi}_{\text{soft}}$ from various sources and to present them in a form that is useful for informative map construction. Cassettari (1993), e.g., maintained that in the context of an integrated geoinformation management system, the process of converting uncertain data to information adds extra values to the original data. The anxiety generated by limited data and uncertain knowledge is poetically expressed by Tulku (1990):

> When the time is short, space is limited, and knowledge is uncertain, where can we turn?

Fortunately, in the case of *TGIS* it turns out that many soft data types commonly available in practical situations can be expressed mathematically as a function of the probability law of $\boldsymbol{\chi}_{\text{soft}}$, i.e.,

$$\Xi_{\mathcal{S}}(\boldsymbol{\chi}_{\text{soft}}) = \text{Funct.}\ \{\text{Prob}[\boldsymbol{x}_{\text{soft}}], \boldsymbol{\chi}_{\text{soft}}\} \tag{3.17}$$

As is usually the case with many mathematical formulas, the meaning of Eq. 3.17 is better understood with the help of a few examples.

Example 3.9. In Table 3.1, Eq. 3.17 is expressed in terms of probability functions of various forms. In particular,

- The $\boldsymbol{\chi}_{\text{soft}}$ often represents *interval* soft data, i.e., $\chi_i \in I_i = [l_i, u_i]$, $i = m_h + 1, \ldots, m$ (where l_i and u_i are the lower and upper bounds of the intervals I_i; see Eq. 3.18 in Table 3.1).
- Other kinds of $\boldsymbol{\chi}_{\text{soft}}$ may have the *probabilistic* form of Eq. 3.19, where $F_{\mathcal{S}}$ is the cumulative distribution function (cdf) constructed on the basis of the $\mathcal{S}$-*KB*.
- Also, $\boldsymbol{\chi}_{\text{soft}}$ may be expressed by means of Eq. 3.20, where the function h(·) represents an empirical chart, model, etc.
- Finally, Eq. 3.21 corresponds to the case in which soft (probabilistic) data are also available at the estimation points $\boldsymbol{p}_k$ themselves.

Table 3.1. Examples of soft data

	$\Xi_{\mathcal{S}}(\boldsymbol{\chi}_{\text{soft}})$	
$\text{Prob}[\boldsymbol{x}_{\text{soft}} \in I] \quad I = (I_{mh}, \ldots, I_m)$	$\boldsymbol{\chi}_{\text{soft}}$	(3.18)
$\text{Prob}[\boldsymbol{x}_{\text{soft}} \leq \boldsymbol{\chi}_{\text{soft}}] = F_{\mathcal{S}}(\boldsymbol{\chi}_{\text{soft}})$	$F_{\mathcal{S}}(\boldsymbol{\chi}_{\text{soft}})$	(3.19)
$\text{Prob}[h(\boldsymbol{x}_{\text{soft}}) \leq \boldsymbol{\chi}] = F_{\mathcal{S}}(\boldsymbol{\chi}, h)$	$F_{\mathcal{S}}(\boldsymbol{\chi}, h)$	(3.20)
$\text{Prob}[\boldsymbol{x}_{\text{soft}} \leq \boldsymbol{\chi}_{\text{soft}}, x_k \leq \chi_k] = F_{\mathcal{S}}(\boldsymbol{\chi}_{\text{soft}}, \chi_k)$	$F_{\mathcal{S}}(\boldsymbol{\chi}_{\text{soft}}, \chi_k)$	(3.21)

Various kinds of uncertain physical data – contaminant concentrations, facies categories, petrophysical properties, soil moisture, geologic images, and seismic data – can be expressed in terms of Eqs. 3.18 through 3.20. As we shall see later in the book (Sect. 4.2.5), Eq. 3.21 represents a desirable situation, also referred to as space/time *filtering*, which can improve considerably the predictive properties of the space/time maps obtained by *BME* modelling. ▪

Example 3.10. Most *TGIS* experts are familiar with the fact that, interval soft data of the form of Eq. 3.18 in Table 3.1 arise naturally in applications related to classification problems. These kinds of problems appear in a variety of scientific and engineering disciplines. Interval data of the sand, silt or clay contents of the various soil types of a geographical region can be derived from the soil map of that region using the relevant texture classification triangle (see e.g., the U.S. Department of Agriculture texture triangle in Fig. 3.6). An interesting case study of soil texture mapping based on the Belgian texture classification triangle is discussed in D'Or et al. (2001). Furthermore, nitrate concentrations can be calculated with considerable accuracy using laboratory procedures which are, though, both lengthy and costly. This often leads to a situation in which a limited number of hard measurements are available, whereas a large amount of soft data already exists or can be easily generated. For instance, since the reaction level between nitrate concentration and a color-strip paper depends on the concentration level, a range of possible nitrate concentrations can be calculated on the basis of the color scale. ▪

Example 3.11. The speed and flexibility of a *GPS* (Global Positioning System) make it ideal for generating data in a large variety of *TGIS* applications. However, the *GPS* readings are often less than perfectly accurate. Possible error sources include faulty clocks, receiver fuzziness, ephemeris variations, multipath "ghosting", and the not-quite-constant speed of light (Steede-Terry 2000). As a result, these sources of *GPS* uncertainty need to be modelled with the help of the soft data equations of Table 3.1. ▪

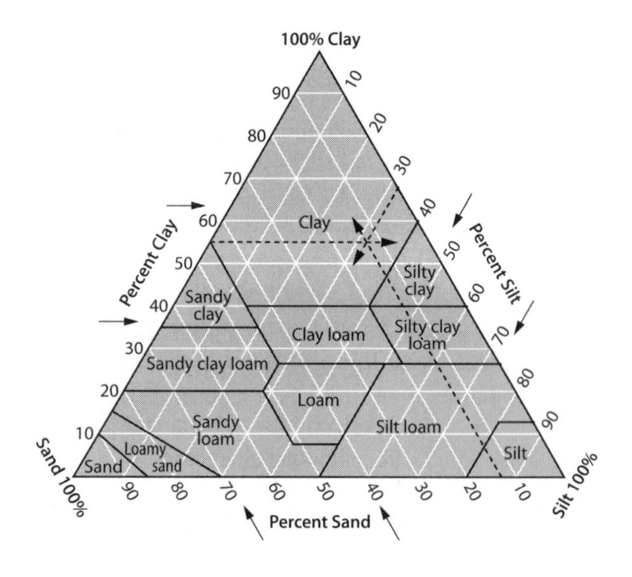

Fig. 3.6. The USDA soil texture classification triangle (after Krumhardt and Wirth 1999)

Most specialists would agree that a major concern of soft data processing in *TGIS* is related to *data quality*. The issue has been appropriately emphasized by Heuvelink (1998), who notices that no map stored in a *TGIS* is error-free. Error contamination is an unavoidable consequence of the lengthy procedure the data collected in the field usually undergo, including classification, interpretation, estimation, and storage. Thus, errors are due to measurement inadequacies, interpolation schemes, boundary determination, reproduction, data transfer and entry, etc. (Muller 1987; Goodchild 1989; Bolstad et al. 1990; Burrough and McDonnell 1998).

Example 3.12. Soft probability data in the form of Eq. 3.19 commonly arise in situations in which there is a measurement error in the laboratory data of soil properties (such as shear strength, soil moisture, or porosity; Christakos 1980). Consider, e.g., an empirical relationship of the form

$$Y(\boldsymbol{p}) = X(\boldsymbol{p}) + W(\boldsymbol{p}) \tag{3.22}$$

where $X(\boldsymbol{p})$, $Y(\boldsymbol{p})$, and $W(\boldsymbol{p})$ are, respectively, the actual (but unknown) strength, the measured shear strength, and the associated measurement error. The $W(\boldsymbol{p})$ is, of course, unknown, but we usually have an idea about the distribution of its possible values. Assume, e.g., that our knowledge of $W(\boldsymbol{p})$ is uncertain and is expressed by the probability function $\text{Prob}[\boldsymbol{w}_{\text{soft}} \leq \boldsymbol{\omega}] = F_W(\boldsymbol{\omega})$. Then, the following soft (probabilistic) data about $X(\boldsymbol{p})$ are obtained, $\text{Prob}[\boldsymbol{x}_{\text{soft}} \leq \boldsymbol{\zeta}] = 1 - F_W(\boldsymbol{\psi} - \boldsymbol{\zeta})$. ▪

Soft data representing the views of decision makers (preferences, priorities, intuitions, judgments, surveys, etc.) could also be used in certain cases, because social and political issues count when *TGIS* are implemented in the decision-making process. In addition, useful although uncertain *KB* may be available in the form of vague statements and qualitative assessments (e.g., "somehow cold", "rather tall", and "to a certain degree the point is inside the region in question"). Proper quantification of such soft data sometimes involves *fuzzy logic* techniques which focus on highly interpretable representation of human application-domain knowledge (Klir and Yuan 1995; Burrough and Frank 1996).

Example 3.13. In some *TGIS* applications, fuzzy techniques have been successfully used to describe geographical regions with imprecisely defined boundaries (e.g., Robinson 1988), to measure uncertainties about the geographical data and decision rules, or to represent imprecise concepts (such as the concept of spatial accessibility discussed in Malczewski 1999). ▪

In most applications, the *TGIS* specialist will need to integrate substantially different data sources (hard and soft). As discussed in Sheppard et al. (1999), e.g., the *TGIS* should allow the integration of the different data sources utilized in environmental equity analysis, such as data on locations of hazardous sites (e.g., toxic release inventory sites, and Superfund sites), and the various population characteristics (e.g., race, age, and social status). Christakos and Serre (2000a) used a *TGIS* that combined hard and soft data on mortality and temperature in order to calculate exposure-health effect associations. Also, Christakos and Serre (2000b) utilized a *BME* method which incorporated hard with soft particulate matter data, thus leading to improved maps of the particulate matter distribution across space and time. Furthermore, the integrated use of *TGIS* operations with a variety of *GPS*-generated datasets can make a huge difference in one's efforts to discover and understand the life cycle of a disease. Steede-Terry (2000), e.g., discusses a case-study in which the life cycle of leishmaniasis in the state of Texas was successfully tracked by using *TGIS/GPS* to plot maps showing the locations of hosts (rats) and the means of transmission (sand flies); then these maps were compared with the distribution of people infected with the disease across space and time.

When dealing with the potential use of soft data in *TGIS* applications, it is often useful to define some kind of an *efficiency index*, which adequately quantifies the advantage of incorporating these soft data into the space/time mapping process. Chapter 6 introduces an efficiency index in terms of the expected mapping accuracy (estimation variances). Also, some numerical illustrations related to this important issue are presented in Sect. 6.4.

3.3.2 The Effect of Soft Data on the Calculation of the Space/Time Correlation Functions

In view of the preceding discussion on the various kinds of uncertain knowledge considered by *TGIS* in practice, the effect of soft data on the calculation of space/time correlation functions is a substantial issue that typically arises in many geographotemporal studies. Preferably, the space/time correlation functions should be calculated on the basis of physical laws, phenomenological relations, etc. (see Sect. 3.2 above). In this way, a series of estimation problems can be avoided (Christakos 2000). If, however, it is necessary to calculate the space/time correlation functions from the existing datasets (since no laws or other sources of information are available in the situation of interest), the fact that these datasets are soft should be taken into account. *Modern Spatiotemporal Geostatistics* provides the necessary tools for such calculations, and an example is discussed next.

Example 3.14. Assume that a soft dataset χ_{soft} is available for a spatially homogeneous random field in which all the data have interval forms, i.e., $\chi_i \in I_i = [l_i, u_i], i = 1, \ldots, m$ (where l_i and u_i are the lower and upper bounds of the intervals I_i; see Eq. 3.18 in Table 3.1). Then, it can be shown that the interval-based (centered) covariance between the two points s_i and s_{i+1} at a distance $h_1 = |s_{i+1} - s_i|$ apart is given by

$$c_{x,I}(s_i, s_{i+1}) = \frac{1}{m}\left[\sum_{i=1}^{m} v_i v_{i+1} - \frac{1}{m}\left(\sum_{i=1}^{m} v_i\right)^2\right] \tag{3.23}$$

where $v_i = ½(u_i + l_i)$, and $\mu_I = {}^1\!/_m \sum_{i=1}^{m} v_i$ is a mean value. I.e., in this case we get the required results by using the value v_i at the center of the corresponding interval I_i associated with each point s_i ($i = 1, \ldots, m$). It must be emphasized, however, that while Eq. 3.23 is equivalent to the interval-based covariance, it may, nevertheless, provide a rather biased estimate of the actual covariance, especially when the widths of the interval data are large. Furthermore, using the v_i values as hard data may cause considerable problems in subsequent stages of space/time mapping (see e.g., the soft data "hardening" trick in Sect. 6.5.1). ▪

The analysis of the previous example can be easily extended to other forms of soft probability data. Next, we conclude this chapter with a brief discussion of a series of issues related to the efficient accommodation of the knowledge needs which arise in the context of today's *TGIS*.

3.4 Accommodating Knowledge Needs

3.4.1 Knowledge Classification

In principle, all kinds of geographical, physical, ecological, epidemiological, economical, social, etc. data can be used in a *TGIS*. The various types of *KB* considered in the previous sections as well as the degrees of uncertainty associated with each type of *KB* require a careful investigation. The comprehensive knowledge needs of *TGIS* analysis may be classified as follows:

a *Available* knowledge needs (i.e., relevant *KB* that are already available to the *TGIS* specialist).
b *Recognized but unavailable* knowledge needs (i.e., *KB* recognized as important, but not available to the *TGIS* specialist).
c *Unrecognized* knowledge needs (i.e., salient *KB* that is not recognized as such by the *TGIS* specialist).

Modern *TGIS* offer a systematic effort aiming at accommodating such needs. The two major *KB* (general and case-specific) discussed in previous sections play a very important role in a realistic *TGIS* study. Indeed, the quality of a map generated by a *TGIS* depends heavily on the availability of high-quality physical knowledge, as represented in the above *KB*. A mathematically sophisticated *TGIS* is of limited use if it is not "fed" with the appropriate *KB*. Poorly developed and processed *KB* can compromise the quality of decision making. Ideally, in practice a *TGIS* specialist would prefer that

KB are systematically organized and accessible

so that they can be processed electronically and interpreted by the computer. Therefore, it would be very useful if mechanisms are developed to facilitate the collection and electronic dissemination of the *KB* discussed in the preceding sections. Naturally, these mechanisms will need to go beyond the current standard of printed journals or the way that experts transfer their experience and expertise to others. Certainly, the significance of the *KB* in a specific situation depends on the way the educated mind of the *TGIS* specialist extracts meaning from these *KB*. There is a huge difference between simply having access to information and having the necessary critical thinking skills to interpret it in an adequate manner.

3.4.2 Model Building and Reality Check

In some respects, model building is viewed as an effective approach for formulating thinking, posing sharp questions, and discovering knowledge. Therefore, theoretical and computational models are used with increasingly frequency by *TGIS* in various scientific disciplines. Models need to be credible in terms of the observables they specify and the observables' link to the geographotemporal processes under consideration. In such a context, the two *golden rules* of *TGIS*-based knowledge synthesis could be summarized as follows:

Testing that all the underlying model assumptions are satisfied by the real-world data

and

Applying site-specific control to theoretical and computational models.

Indeed, the *assumptions* of the (theoretical and computational) models must be carefully tested vs. the real-world data, before we use them to study a *TGIS* application[8]. When this is not possible at the initial stage of the study, we should at least keep track of the untested assumptions. Models should not be viewed as dogmas, forcing the real-world observations into compliance with them. Instead, it may turn out that, in some cases modifying the model assumptions in light of the real-world data moves knowledge of the system forward. Also, *site-specific control* of these models is necessary to make sure that the relevant environmental influences (geological, hydrological, atmospheric, etc.) are taken into account by the model solutions and predictions. It is surprising how often these two rules are neglected by some *TGIS* practitioners in various scientific disciplines, thus leading to the mishandling and misjudgment of a potentially very powerful tool.

On a relevant note, most *TGIS* specialists are familiar with a salient issue that concerns the decision of how *complex* the models should be that they can expect to use with reasonable efficiency in real-world applications. This is by no means a trivial issue. While a simple model may leave out important components of the natural phenomenon, a model that is too complex may face the problem of inadequate site-specific data to estimate its various parameters. On the other hand, a simple model of a life-support system may underestimate the prediction uncertainty of that system, whereas a more complex model could offer a better representation of reality. The situation has been studied to considerable depth by various authors (see e.g., Linhart and Zucchini 1986; Hilborn and Mangel 1997). By way of a summary, it is reasonable to expect that, when a specialist makes a decision regarding the complexity of the model to use in a *TGIS*, he should take into account two crucial factors:

a The *availability* of adequate site-specific data; and
b The intended *use* of the model by the *TGIS* practitioner.

An optimal model for the situation should obtain a proper balance between these two crucial factors. Also, due to changes occurring across space and time, any such model should take into account the contextual settings of the databases.

[8] Assumptions that do not hold give rise to models that operate only in an imaginary world, and are of limited use in the study of real-world situations. For a model to be useful, its assumptions must match at least certain significant aspects of reality.

Chapter 4

Spatiotemporal Mapping

If you put something simply, you are at the mercy of those who understand neither the subject nor simplicity.
E. De Bono

4.1 A Formulation of the Spatiotemporal Mapping Problem

For the *TGIS* purposes of this book, we start the chapter with a general formulation of the *spatiotemporal mapping* problem that could be of interest to a variety of scientific disciplines (see also Fig. 4.1):

- Consider a natural, epidemiological or social field $X(\boldsymbol{p})$ whose distribution across space and time is characterized by a general $\mathcal{G}$-*KB* and a specificatory $\mathcal{S}$-*KB* at a set of data points $\boldsymbol{p}_{\text{data}}$.
- An *S/TRF* estimator $\hat{X}(\boldsymbol{p})$ is sought that provides estimates of the actual (but unknown) $X(\boldsymbol{p})$ values at an arbitrary set of mapping points $\boldsymbol{p}_k$ in the space/time domain.

To solve the spatiotemporal mapping problem above, one needs to construct a mathematical operator which uses all relevant *KB* (i.e., including but not limited to the observations available at the data points) in order to generate maps that approximate the spatiotemporal distribution of $X(\boldsymbol{p})$ at unexplored or poorly explored space/time domains represented by the mapping points.

Comment 4.1. In most applications, the mapping points lie on the nodes of a space/time grid. Depending on the situation, the spatial component of the grid may cover a plane surface or a curved surface (see also Sect. 2.1).

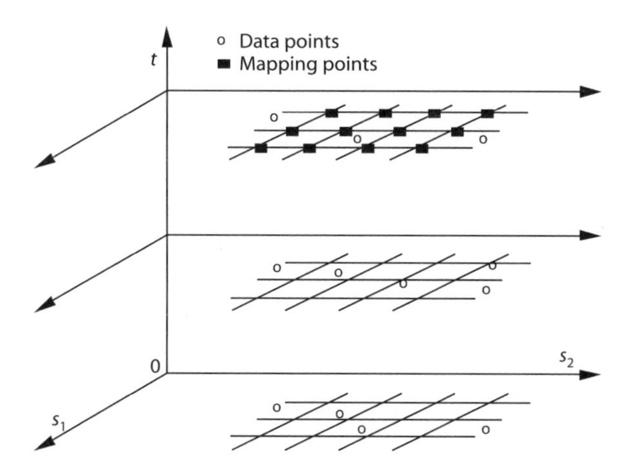

Fig. 4.1. An illustration of the spatiotemporal mapping problem in *TGIS*

It is worth noticing that spatiotemporal mapping may be concerned with single-point or multi-point space/time estimation, as follows:

- *Single-point* analysis focuses on a single estimate $\hat{\chi}_k$ of the field value χ_k at a point $\boldsymbol{p}_k$, at a time. In this case the vectors $\boldsymbol{x}_{\text{map}}$, $\boldsymbol{p}_{\text{map}}$, and $\boldsymbol{\chi}_{\text{map}}$ are written as,

$$\left.\begin{aligned} \boldsymbol{x}_{\text{map}} &= (x_1, \ldots, x_m, x_k) \\ \boldsymbol{p}_{\text{map}} &= (p_1, \ldots, p_m, p_k) \\ \boldsymbol{\chi}_{\text{map}} &= (\chi_1, \ldots, \chi_m, \chi_k) \\ \hat{\chi}_k & \end{aligned}\right\} \quad (4.1)$$

- *Multi-point* analysis deals with the estimates $\hat{\boldsymbol{\chi}}_k = (\hat{\chi}_{k1}, \ldots, \hat{\chi}_{k\rho})$ of the field values $\boldsymbol{\chi}_k = (\chi_{k1}, \ldots, \chi_{k\rho})$ at several points $\boldsymbol{p}_{k\ell}$ ($\ell = 1, \ldots, \rho$) simultaneously. Now the vectors $\boldsymbol{x}_{\text{map}}$, $\boldsymbol{p}_{\text{map}}$, and $\boldsymbol{\chi}_{\text{map}}$ become,

$$\left.\begin{aligned} \boldsymbol{x}_{\text{map}} &= (x_1, \ldots, x_m, x_{k_1}, \ldots, x_{k_\rho}) \\ \boldsymbol{p}_{\text{map}} &= (p_1, \ldots, p_m, p_{k_1}, \ldots, p_{k_\rho}) \\ \boldsymbol{\chi}_{\text{map}} &= (\chi_1, \ldots, \chi_m, \boldsymbol{\chi}_k) \\ \boldsymbol{\chi}_k &= (\chi_{k_1}, \ldots, \chi_{k_\rho}) \\ \hat{\boldsymbol{\chi}}_k &= (\hat{\chi}_{k_1}, \ldots, \hat{\chi}_{k_\rho}) \end{aligned}\right\} \quad (4.2)$$

At first sight, one might be tempted to ask why *deterministic* interpolation techniques (polynomial fitting, spline functions, etc.) should not be used in scientific space/time mapping. Unfortunately, such techniques are clearly inadequate, since the data usually available are not merely evaluations of some analytic function, and the spatiotemporal analysis of natural fields is much more than classical interpolation. Instead, the natural fields are governed by physical laws and are related to each other. Furthermore, the data may involve errors and other sources of uncertainty that are spatiotemporally correlated with each other as well as with different fields (e.g., Daley 1991). At many points the data are available in a "soft" form rather than in a perfectly "hard" form (as described in Sect. 3.3). And, several kinds of general knowledge may be available at the prior stage of the analysis (as in Sect. 3.2). Deterministic interpolation techniques cannot take into account such essential aspects of *TGIS* mapping and, thus, their usefulness in real-world situations is rather limited.

Some earlier *statistical* interpolation techniques were based on the so-called "naive empiricism" approach[1]. At the heart of the naive empiricism techniques is the argument

Res ipsa testit.

[1] The problematic character of "naive empiricism" is eloquently articulated by Popper (1962), and by Wang (1993).

(I.e., facts speak for themselves.) To this group belong several statistical curve-fitting and surface-fitting techniques (Bennett 1979) as well as neural network techniques (Cherkassky and Mulier 1998). Seeking to reduce problems to forms to which these techniques are applicable often generates data fits that merely satisfy statistical criteria (e.g., most regression techniques use a minimum mean squared error criterion), possibly at the expense of the attainable scientific realism that is required for a physically meaningful *TGIS* analysis. By avoiding theoretical modelling and relying solely on the hard data, mapping techniques based on naive empiricism have limited predictive or extrapolation power[2]. Certain of the techniques (e.g., neural networks) have serious interpretability problems. Moreover, by trying to find quick and easy solutions, naive empiricists are often distracted from dealing with the deeper problems underlying many real-world situations. Today, naive empiricism appears to be at best a primitive stage through which investigations may pass on their way to becoming mature sciences (e.g., Newton 1997). The dominant view of scientific development, known as the "hypothetico-deductive" approach, emphasizes hypothesis and prediction, in addition to explanation[3]. Other mapping techniques (including the Wiener-Kolmogorov interpolator and the Kalman-Bucy filter) will be visited in Sect. 4.3, following the discussion of the *BME* approach in Sect. 4.2 below. It is our hope that new insight can be gained about these well-known techniques, in the light of modern epistemic analysis.

We conclude this introductory section by noticing that, faced with a variety of mapping techniques, *TGIS* specialists should always exercise the necessary judgment in choosing the one technique that is the most appropriate for the job at hand. There are meaningful criteria for such a purpose, such as

- The assumptions made by the technique should be consistent with the physical characteristics of the phenomenon being mapped.
- The domain of the technique's applicability is well defined.
- The data and computational resources should exist for the proper application of the technique, etc.

Burrough (1986) brings this important issue to our attention by plausibly suggesting that

> It is unwise to throw one's data into the first available interpolation technique without carefully considering how the results will be affected by the assumptions inherent in the methods.

With this last note in mind, we proceed with a brief introduction of the *BME* modelling and mapping approach of *Modern Spatiotemporal Geostatistics.*

[2] This may not be a surprise, after all. Naive empiricism, which flurished in the geological works of the late 19th century, continued to play a major role in 20th century geology. These works focused on explanation based on past evidence, whereas little attention was paid to prediction. Geologists' concern was the past (about which data collection was usually a sure thing), and not the future (about which data gathering was clearly inconceivable). As Oreskes (2000) points out, in the geologists' mind prevailing was the dictum that, "the future, by definition, involved the unobservable – the inaccessible – and therefore pushed one beyond the realm of inductive science". Any tangible remains of the future being plausibly absent and by lacking any sound theoretical model of it, one has serious difficulties to talk about its prediction.

[3] In recent years, the hypothetico-deductive method has become dominant to the point of being accepted by most geoscientists, as well (see also Chap. 8, later).

4.2 Formal *BME* Analysis and Mapping

Through spatiotemporal mapping one can achieve systematically and with the support of the appropriate physical *KB* an improved understanding of a natural phenomenon or a situation. Kant maintained that:

> Reason has insight only into that which it produces after a plan of its own.

A remarkable epistemic implication of this statement is that the design of any experimental investigation is, essentially, the specification *a priori* of a model in terms of which the experimental results will be described *a posteriori*[4]. This approach leads to the three main stages of knowledge acquisition, processing, and integration.

The first two stages are, the *prior* (or *structural*) stage, and the *meta-prior* stage. To each one of these two stages there corresponds a *KB* representing the structural aspects (general *𝒢-KB*, for the prior stage) and the site-specific aspects (specificatory *𝒮-KB*, for the meta-prior stage) of the situation of interest. The consideration of the *𝒢-KB* at the prior stage reflects the fact that in physical applications we do not start with blank ignorance of what a natural process represents. Instead, we often possess a rather vague or merely practical grasp of the situation, and we seek to pass to an explicit and reasoned grasp of what the natural phenomenon is. In the words of Armstrong (1983):

> We do not go from black night to daylight, but from twilight to daylight.

This passage from the general to the specific includes the consideration of the site data (*𝒮-KB*) at the meta-prior stage. The third stage is the *integration* stage (also called the *posterior* stage), in which the previous two stages are integrated, thus leading to the final outcome of the analysis. This stage makes possible the *horizontal integration* of disparate scientific disciplines and, as a consequence, it generates an improved picture of reality. If, e.g., it succeeds to process a variety of general and specificatory *KB*, the *TGIS* could bring together several scientific disciplines which are all relevant to the problem under study. The implication of all this is that, *TGIS* could become a vital component of an *interdisciplinary* approach to complex problems of life support sciences.

The preceding lines of thought constitute the philosophical underpinnings of the *BME* approach. As is discussed in Christakos (1998a, 1998b, 2000), the conceptual *BME* approach has two distinct parts:

i. A *formal* component focusing on mathematical structure, logical process, and theoretical representations, combined with
ii. An *interpretive* component concerned with applying the formal part in real-world situations, including the physical meaning of mathematical structure, specific observation methods and connections to other empirical phenomena.

Interpretation issues are relevant when one needs to establish relationships (also called, *correspondence* or *operational* rules) between the life support-related fields and

---

[4] The reader may recall that epistemic analysis is concerned with the theory of knowledge (i.e., the acquisition, structure, processing, as well as the criteria of knowledge).

the formal mathematics which describe them, to measure and test the formal structure, or to justify certain methodological steps. The text that follows concentrates on the formal *BME* approach, which is based on a systematic and rigorous framework. The framework accounts for various forms of life support-related *KB*, thus improving considerably the scientific content and accuracy of the spatiotemporal map. Interpretive *BME* will be the subject of Chap. 5.

4.2.1 The Basic *BME* Procedure

There are six distinct steps in formal *BME* analysis and mapping. These steps are briefly summarized below[5]:

Step 1. In view of the general $\mathcal{G}$-*KB* considered in the particular *TGIS* application (including physical laws, scientific theories, and multiple-point statistics), formulate the corresponding stochastic equations of the $\overline{h_\alpha}$ and $\overline{g_\alpha}$ moments – see Eqs. 3.2. The underlying g_α functions ($\alpha = 1, \ldots, N$) should be properly chosen so that they represent all the physical *KB* available, and the total number N of Eqs. 3.2 must be such that moments involving all the space/time points of the map, $\boldsymbol{p}_i \in \boldsymbol{p}_{\text{map}}$, are included[6].

Step 2. Assume that the $\mathcal{G}$-based pdf model $f_\mathcal{G}$ has the following form

$$f_\mathcal{G}(\boldsymbol{\chi}_{\text{map}}) = e^{\mu_0 + \boldsymbol{\mu}^T \boldsymbol{g}} \tag{4.3}$$

where $\boldsymbol{g} = \{g_\alpha;\ \alpha = 1, \ldots, N\}$ is the vector of g_α-functions defined in *Step* 1 above, $\boldsymbol{\mu} = \{\mu_\alpha;\ \alpha = 1, \ldots, N\}$ is a vector of coefficients associated with $\boldsymbol{g}$, and μ_0 is a coefficient that accounts for the normalization constraint $\overline{g_0} = 1$. The elements of the $\boldsymbol{\mu}$ are functions of the space/time coordinates and will be determined in the following *Step* 3.

Step 3. Substitute Eq. 4.3 into Eq. 3.2 and solve for the coefficients $\boldsymbol{\mu}$. Then, insert the coefficients $\boldsymbol{\mu}$ back into Eq. 4.3 to find the exact analytical form of the $\mathcal{G}$-based pdf $f_\mathcal{G}$ of the map.

Step 4. In light of the specificatory $\mathcal{S}$-*KB* available, the $\mathcal{G}$-based pdf $f_\mathcal{G}$ is updated by means of *Bayesian conditionalization* leading to the *BME* pdf for the map as follows

$$f_\mathcal{K}^{bc}(\boldsymbol{\chi}_k) = A^{-1} \int_D d\Xi_\mathcal{S}(\boldsymbol{\chi}_{\text{soft}}) f_\mathcal{G}(\boldsymbol{\chi}_{\text{map}}) \tag{4.4}$$

where the superscript bc means Bayesian conditional (see also Chap. 5), the subscript $\mathcal{K} = \mathcal{G} \cup \mathcal{S}$ denotes the total knowledge considered, $A = \int_D d\Xi_\mathcal{S}(\boldsymbol{\chi}_{\text{soft}}) f_\mathcal{G}(\boldsymbol{\chi}_{\text{data}})$ is a normalization constant, and the integrand $\Xi_\mathcal{S}$ was defined in Sect. 3.3 (its form depends on the corresponding soft data available, and so does the form of the domain D; see

[5] For a detailed analysis of each step, see Christakos (2000).

[6] Examples of g_α-functions were discussed in Sect. 3.2 above; several more will be presented in Chap. 6.

Table 4.1. Examples of D

S	D
Eq. 3.18	I
Eq. 3.19	I
Eq. 3.21	$I \cup I_k,\ I_k = (I_{k_1}, \ldots, I_{k\ell})$

Table 4.2. Examples of space/time estimates $\hat{\chi}_k$

BMEmode	$\hat{\chi}_{k,\text{mode}} : \max_{\chi_k} f_{\mathcal{K}}^{bc}(\chi_k)$	(4.5)
BMEmean	$\hat{\chi}_{k,\text{mean}} = \int d\chi_k \chi_k f_{\mathcal{K}}^{bc}(\chi_k)$	(4.6)

Table 4.1)[7]. The last case in Table 4.1 corresponds to space/time *filtering*, i.e., the case where soft data are available at the estimation points themselves (see also Sect. 4.2.4 below).

Step 5. Depending on the goals of the study, select the appropriate space/time estimates $\hat{\chi}_k$ (see Table 4.2). The *BMEmode* estimate represents the most probable value, whereas the *BMEmean* minimizes the mean square estimation error. Several other estimates could be derived so that an expected objective function $\overline{\Phi(x_k)}$ is optimized with respect to these estimates, where the form of the function Φ as well as the optimization conditions depend on the characteristics of the situation considered.

Step 6. Because of the inherent randomness of the distribution of the field values and the inaccuracies of the physical data, one needs to use the *BME* pdf 4.4 to obtain an uncertainty assessment associated with the space/time estimate. Let us first consider the single-point estimation case, see Eq. 4.1. In general, for a selected confidence level η, a *BME* confidence set Φ_η can be defined as follows:

$$\chi_k \in [\chi_{k,l},\ \chi_{k,u}] : \qquad P[\chi_{k,l} \le \chi_k \le \chi_{k,u}] = \eta \tag{4.7}$$

where $\chi_{k,l}$ and $\chi_{k,u}$ are the lower and upper bounds of the confidence set, respectively. The choice of these bounds varies. For instance, a popular measure of space/time estimation accuracy is the standard deviation

$$\sigma_{k|\mathcal{K}}^{bc} = [\int d\chi_k (\chi_k - \hat{\chi}_{k,\text{mean}})^2 f_{\mathcal{K}}^{bc}(\chi_k)]^{1/2} \tag{4.8}$$

of the pdf $f_{\mathcal{K}}^{bc}$. Note that, unlike the classical Kriging error standard deviation which is independent of the data values (and, as a consequence, it has been the subject of some

[7] Note that in some more complicated cases of site-specific *KB*, the right hand side of Eq. 4.4 may be multiplied by a parameter B, see Christakos (2000). For the applications considered in this volume $B = 1$.

Table 4.3. Uncertainty measures for *TGIS*

Symmetric *pdf*	$\left.\begin{array}{l}\chi_{k,\ell} = \hat{\chi}_{k,\mathrm{mean}} - \alpha\sigma^{bc}_{k\mid\mathcal{K}} \\ \chi_{k,u} = \hat{\chi}_{k,\mathrm{mean}} + \alpha\sigma^{bc}_{k\mid\mathcal{K}}\end{array}\right\}$	(4.9)
.Asymmetric *pdf* (single maximum)	$\left.\begin{array}{l}\chi_{k,\ell} = \hat{\chi}_{k,\mathrm{mean}} - \theta_{k,\ell} \\ \chi_{k,u} = \hat{\chi}_{k,\mathrm{mean}} + \theta_{k,u}\end{array}\right\}$	(4.10)

criticism among geostatisticians; Goovaerts 1997), $\sigma^{bc}_{k|\mathcal{K}}$ depends on the specific dataset considered. In the case of a *symmetric pdf* (that is characterized by a centrality and dispersion parameters), the usual confidence interval can be defined in terms of $\sigma^{bc}_{k|\mathcal{K}}$; see Eq. 4.9 of Table 4.3, where the coefficient α depends on the choice of η (e.g., for a $\eta = 95\%$ confidence the corresponding $\alpha = 1.96$ for a Gaussian distribution). In the case of an *asymmetric pdf* with a single maximum, a confident width can be defined, see Eq. 4.10 of Table 4.3, where the $\theta_{k,l}$ and $\theta_{k,u}$ are such that $f^{bc}_{\mathcal{K}}(\chi_{k,l}) = f^{bc}_{\mathcal{K}}(\chi_{k,u})$. Equation 4.7 can be generalized to the multi-point estimation domain, see Eq. 4.2, i.e.,

$$\boldsymbol{\chi}_k \in \Phi_\eta : \qquad P[\boldsymbol{\chi}_k \in \Phi_\eta] = \eta \tag{4.11}$$

in which case the interval $[\chi_{k,l}, \chi_{k,u}]$ of Eq. 4.7 above has been replaced by the multidimensional set Φ_η.

We conclude this section by noticing that the epistemic underpinnings of the six *BME* steps above can be summarized as follows:

Keep the derived maps consistent with the site-specific data while exactly satisfying the constraints arising from the general knowledge.

The above are very fruitful epistemic principles which can be put to important use. Due to their importance in *TGIS* methodology, the epistemic principles of the *BME* approach are further explored in Chap. 5.

4.2.2 The Advantage of Composite Space/Time Mapping

In general, composite space/time mapping has considerable advantages over purely spatial mapping. These advantages (theoretical and computational) are discussed, in Christakos and Vyas (1998), Christakos and Serre (2000), Christakos et al. (2001). On the basis of real-world data these references demonstrate that, by taking into account space/time cross-correlations, *BME* produces maps that are more accurate and informative than the maps of statistical analysis in which the spatial fields are estimated separately from the temporal processes. The California case-study is briefly discussed in the following example. Additional numerical examples are considered in Chap. 6. The reader can study these examples with the help of the *BMElib* included in this book.

Example 4.1.[8] We seek to generate spatiotemporal maps of the annual particulate matter (PM_{10}) geometric mean in the state of California. PM_{10} refers to particulate matter in the air that has a diameter $d < 10$ μm. It is an air pollutant that is harmful to humans, and the state of California has set a maximum allowable limit of 30 μg m^{-3} for the annual geometric mean in order to protect the population against long-term exposure. The annual geometric mean is defined as the natural anti-log of the average of a set of natural log-transformed daily PM_{10} measurements. One aspect that is particular to the annual geometric mean is that the set of measurements used to calculate this average changes from year to year and from monitoring station to monitoring station (the number of measurements may vary because of missing values, inaccurate readings, shut-down of a monitoring station, etc.). As a result, the calculated values of the annual PM_{10} geometric mean have varying levels of reliability and are considered uncertain (soft) data rather than accurate (hard) data. In order to address this important issue, we characterized the calculated values of the annual geometric mean in terms of probabilistic soft data, and we modelled their variability by means of a spatiotemporal covariance. The probabilistic soft data and covariance used in this example were obtained in Christakos et al. (2001) on the basis of a set of log-transformed PM_{10}, 24-hour averaged measurements collected in California during the 1987–1997 time period. In particular, the probabilistic soft data for the *S/TRF*, $Y(\boldsymbol{s},t)$, representing the annual mean of logPM_{10} was obtained using a regression procedure on log-transformed 24-hour averaged PM_{10} measurements collected at 191 monitoring stations in California during the 1987–1997 time period. This procedure produced the necessary pdf, $f_S(\chi_s)$, representing the soft data of the type defined in Eq. 3.19 at each monitoring station and each year for which sufficient daily PM_{10} measurements were available. Figure 4.2

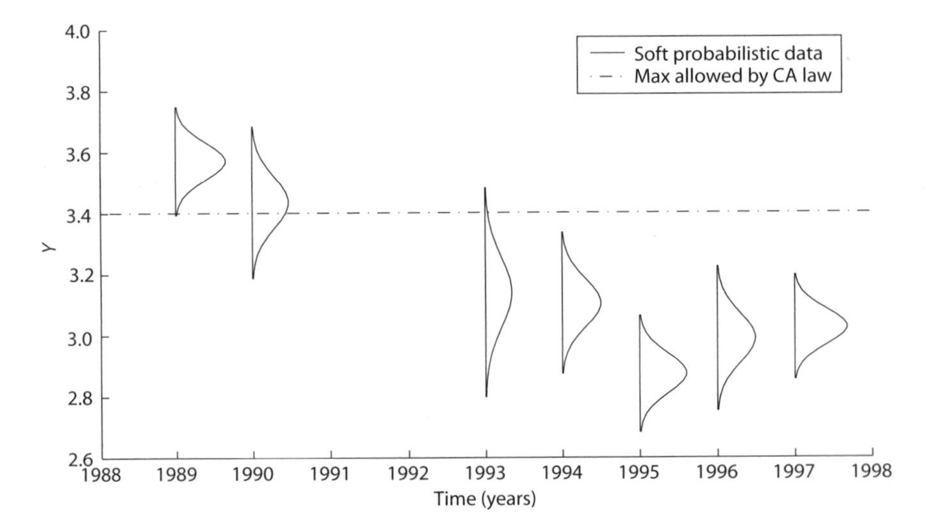

Fig. 4.2. Soft probabilistic data obtained at monitoring station No. 73

[8] The reader can use the routine *ExamplePM10CA.m* to reproduce the results of this example.

shows the soft probabilistic data at monitoring station No. 73. The varying shapes of these soft pdf reflect the varying reliability of the soft data, whereas the missing data correspond to years during which insufficient data were available to calculate the annual $\log PM_{10}$ mean. Also shown with a horizontal line is the maximum permissible limit (i.e., 3.40 = log30) set by the state of California. Figure 4.3 shows the spatiotemporal covariance function used to model the variability of the annual mean of $\log PM_{10}$, $Y(s,t)$. The covariance function has a space/time non-separable shape that consists of the superposition of two exponential models with different spatial and temporal scales, i.e.,

$$c_y(r,\tau) = c_{01}\exp(-3r/a_{s1})\exp(-3\tau/a_{t1}) + c_{02}\exp(-3r/a_{s2})\exp(-3\tau/a_{t2}) \tag{4.12}$$

The first covariance component (with parameters $c_{01} = 0.013$, $a_{s1} = 6$ km, and $a_{t1} = 4$ years) models the small-scale structure of air pollutant fluctuations in space and time, whereas the second component (with parameters $c_{02} = 0.005$, $a_{s2} = 270$ km, and $a_{t2} = 135$ years) represents the large-scale structure of the fluctuations. The large-scale structure may be linked to large meteorological patterns (such as large-scale wind patterns, temperature inversions, etc.), while the small-scale structure is indicative of anthropogenic activities and point source pollution[9]. Using the space/time covariance as general knowledge and the soft probabilistic data as specificatory knowledge, the single point mapping analysis leads to the complete description of the possible values, ψ_k, of the annual mean of $\log PM_{10}$ at the estimation point $\boldsymbol{p}_k$, in terms of the *BME* posterior pdf

$$f_{\mathcal{K}}^{bc}(\psi_k;\boldsymbol{p}_k) = A^{-1}\int_D d\Xi_S e^{\mu_0+\mu^T g} \tag{4.13}$$

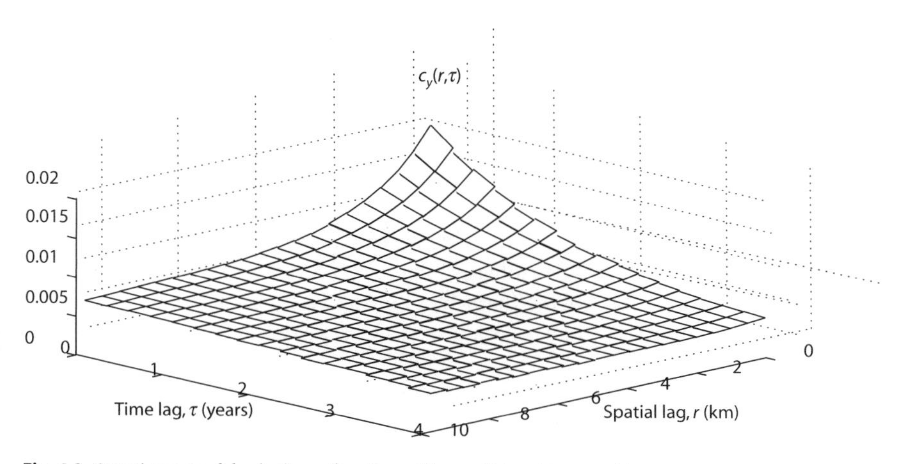

Fig. 4.3. Covariance model $c_y(r,\tau)$ as a function of the spatial and temporal lags

[9] This intepretation is consistent with that of other studies, such as the PM_{10} mapping analysis in Cairo by Serre et al. (2000). Also, note that the shape of the covariance model in Fig. 4.3 can be related to a decomposition of the space/time variability into three random components (i.e., a purely spatial component, a purely temporal component, and an interaction space/time component) as discussed in Bogaert and Christakos (1997).

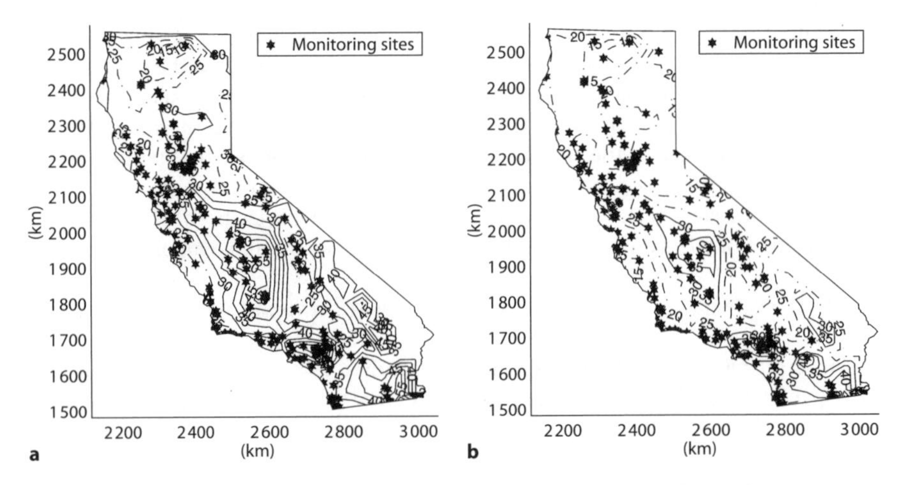

Fig. 4.4. Maps of annual PM_{10} geometric mean in the state of California during the year: **a** 1987; and **b** 1996

where the *S-KB* operator Ξ_S is given by Eq. 3.19. Using *BMElib* we calculated the mean of the posterior pdf (*BMEmean* estimator), and the associated estimation error variance on a grid of points covering the state of California during selected years. The *BMEmean* estimator provides the estimates which minimize the mean squared estimation errors of the annual mean of $\log PM_{10}$. By taking the natural anti-log, the map of the annual geometric mean of PM_{10} was then obtained. For illustration purposes, Fig. 4.4 shows the *BME* maps of the annual PM_{10} geometric mean in California obtained for the years 1987 and 1996. These maps are useful for regulatory and decision-making purposes, since they allow, e.g., to delineate the non-attainment areas for the California maximum permissible limit of 30 μg m^{-3}. The maps were more accurate and informative than those obtained using traditional methods (statistical regression or Kriging), because *BME* rigorously processes probabilistic data, accounts for the composite space/time PM_{10} variability, and offers considerable flexibility in the choice of the appropriate estimator. ▪

BME analysis can take advantage of the existence of a *dynamic* (i.e., time-dependent) physical model to generate improved estimates in time over an area where few or no data are available. Such models include the (baroclinic) primitive equations of atmospheric dynamics, the (unsteady-state) flow equations in subsurface environments, etc. In a sense,

Through the dynamic model-based BME *analysis, information is propagated from well-sampled areas to poorly sampled areas.*

This is called the *global mapping* feature of *BME* because extrapolation can be meaningful beyond the range of observation points. The following example describes an interesting situation in which a spatial observation network is continuously updated in real time – such situations are common in weather forecasting, atmospheric data analysis, and elsewhere.

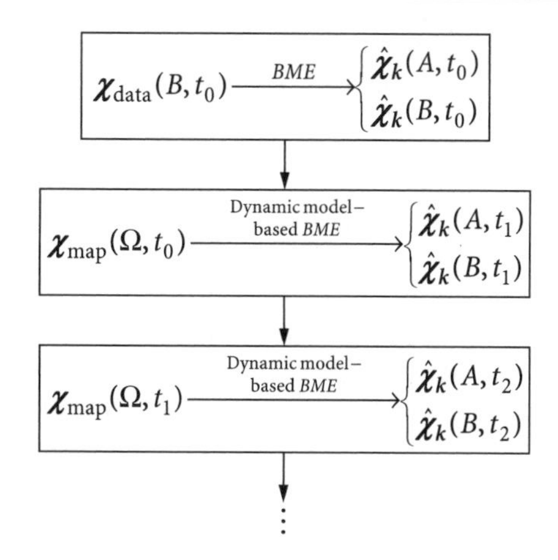

Fig. 4.5. An illustration of the dynamic model-based *BME*

Example 4.2. Consider an area $\Omega = A \cup B$, where A is an unobserved (or very poorly observed) subarea and B is a subarea where spatial measurements are obtained on a rather regular basis in time (e.g., A may be surrounded by B). Figure 4.5 helps us illustrate the global mapping feature. Application of *BME* at the initial observation time t_0, using the data $\chi_{\text{data}}(B,t_0)$ obtained over area B, presumably will give good spatial estimates $\hat{\chi}_k(B,t_0)$ at a set of estimation points p_k in B and rather poor estimates $\hat{\chi}_k(A,t_0)$ at estimation points in A. Clearly, we can write

$$\chi_{\text{map}}(\Omega,t_0) = \chi_{\text{data}}(B,t_0) \cup \hat{\chi}_k(A,t_0) \cup \hat{\chi}_k(B,t_0) \tag{4.14}$$

Next, assume that a dynamic physical model is integrated into the *BME* analysis, thus generating new predictions at a time $t_1 > t_0$ over A and B, i.e. $\hat{\chi}_k(A,t_1)$ and $\hat{\chi}_k(B,t_1)$, respectively. Presumably, $\hat{\chi}_k(A,t_1)$ will be of improved accuracy compared to $\hat{\chi}_k(A,t_0)$. At time t_1 new observations $\chi_{\text{data}}(B,t_1)$ become available. Using

$$\chi_{\text{map}}(\Omega,t_1) = \chi_{\text{data}}(B,t_1) \cup \hat{\chi}_k(A,t_1) \cup \hat{\chi}_k(B,t_1) \tag{4.15}$$

the *BME* approach generates new predictions at time $t_2 > t_1$, where $\hat{\chi}_k(A,t_2)$ is more accurate than $\hat{\chi}_k(A,t_1)$. In this way, *BME* generates successively improved maps for the unobserved area A. ▪

4.2.3 Continuous-Valued Map Reconstruction

Before the computer age, the construction of informative maps depended heavily on the existence of a dense spatial sampling network, since no rigorous and systematic spatial estimation techniques were in use. As is often the case in these situations, there was a major concern about the cost of data collection. A compromise was usually made to obtain dense but moderately accurate measurements, which were cheaper and easier to acquire. As a consequence, in the vast majority of cases very few or no exact measurements were made in order to construct these often goal-oriented maps. Typical examples of this soft data-based mapping are the soil maps. A large number of these maps are available over extended areas which provide the soil type at any spatial location, according to some classification procedures. Since the maps were designed mainly for agricultural purposes, there was no need for a very precise determination of the relevant soil properties (e.g., clay content). Instead, the soil maps could only provide broad information about these properties (e.g., in terms of intervals of values).

Given the situation described above, a crucial question soon arose: "Can we reprocess the soft data-based maps in order to get more accurate information about a continuous-valued soil property, without making any additional measurements?" This kind of accurate information is necessary in a variety of situations. Quantitative environmental models, e.g., which involve a clay content field cannot easily process soft data about this field but, instead, they require the use of exact clay content measurements. Getting new, accurate measurements by means of a dense sampling network would be very costly and time consuming. Thus, a question of great significance is how to get informative estimates of a spatially continuous property at any location in space, using only soft data-based maps. It turns out that the *BME* approach is particularly suited to such a task, due to its inherent ability to process soft data rigorously and efficiently. Numerical illustrations are given in various places of Chap. 6.

4.2.4 Modifications of the *BME* Procedure

A number of theoretical modifications of the step-by-step procedure above are possible. These modifications concern mainly the (structural or prior) *Step* 2 and the (integration or posterior) *Step 4*. Some of these modifications are briefly presented below, but surely the matter is worth further investigation. Before proceeding with our discussion of these modifications, it must be noted that in comparing the current *BME* formulation to alternative formulations resulting from the modifications, the *BME* formulation has certain important virtues: it leads to a simpler yet powerful set of equations, has several appealing and well-understood analytical and computational features, and is epistemically well-grounded.

4.2.4.1 *Fisher Information-Based Prior pdf*

An epistemic justification of Eq. 4.3 in *Step* 2 is based on the choice of the *Shannon* information concept (see discussion in Christakos 2000, and in Chap. 5). The versatility of the *Modern Spatiotemporal Geostatistics* paradigm makes it possible to

Use information measures other than that of Shannon. One of the options available to the TGIS expert is the Fisher *information*

(for a definition of the Fisher information, see e.g., Frieden 1991). It should be noted that, maximization of the epistemic Fisher information is essentially equivalent to minimization of the statistical Fisher information in a space/time mapping context, i.e.

$$\overline{\sum_{i=1}^{m,k}\left[\frac{\partial}{\partial \chi_i}\log f_{\mathcal{G}}(\boldsymbol{\chi}_{map})\right]^2},$$

subject to the $\mathcal{G}$-*KB* available[10]. The Fisher information naturally leads to a non-entropic formulation of the mapping approach and $f_{\mathcal{G}}$ models which are, in general, different than the pdf models of Eq. 4.3. The new geostatistical method is called the *Bayesian Maximum Fisherian* (*BMF*) technique. The *BMF* is illustrated in the following example.

Example 4.3. For simplicity, since our goal is to demonstrate the different mathematical form of the pdf model derived in terms of the Fisher information, we focus on the univariate model $f_{\mathcal{G}}(\chi)$. Assume that the $\mathcal{G}$-*KB* available is expressed in terms of the $g_\alpha(\chi)$-functions ($\alpha = 0, 1, \ldots, N$), as usual. Then, while in the case of the Shannon information the pdf model has the exponential form of Eq. 4.3 above, in the case of the Fisher information the pdf model is the solution of the differential equation

$$\frac{d^2}{d\chi^2}f_{\mathcal{G}}(\chi) - \frac{1}{2f_{\mathcal{G}}}\left[\frac{d}{d\chi}f_{\mathcal{G}}(\chi)\right]^2 - f_{\mathcal{G}}(\chi)\sum_\alpha \mu_\alpha g_\alpha(\chi) = 0 \tag{4.16}$$

The solution of Eq. 4.16 depends on the form of the $g_\alpha(\chi)$-functions, and is obviously a more complicated procedure compared to the exponential Eq. 4.3. Note that *BME* [Eq. 4.3] and *BMF* [Eq. 4.16] yield the same result only in some special cases (e.g., under certain conditions involving the first two statistical moments as the general knowledge $\mathcal{G}$, both information concepts lead to an exponential pdf model). ▪

4.2.4.2
Natural Laws with a Convenient Mathematical Form

Another interesting modification of *Step 2* assumes the direct involvement of the natural law under consideration. More specifically, in some *TGIS* applications,

The relevant natural law may have a convenient form, in which case the prior pdf model could be derived directly from the law.

An example illustrating such a modification is discussed below. In most practical situations, however, the mathematical form of the natural law of interest is considerably complicated, in which case assumption 4.3 leads to a more tractable approach than the law-based modification.

Example 4.4. Assume that the $\mathcal{G}$-*KB* available includes the physical model below

$$\frac{\partial}{\partial t}\boldsymbol{\chi}_{\mathrm{map}} = W(\boldsymbol{\chi}_{\mathrm{map}}) \tag{4.17}$$

at each point $\boldsymbol{p}_i = (\boldsymbol{s}_i, t)$ $(i = 1, \ldots, m, k)$, where $\partial/(\partial t)\chi_i = W_i(\boldsymbol{\chi}_{\mathrm{map}})$, and $\boldsymbol{\chi}^0_{\mathrm{map}}$ are the corresponding initial conditions with known pdf f_0. Let the solutions of Eq. 4.17 be given by

[10] For a more detailed discussion of the epistemic approach to space/time mapping, the reader is referred to the relevant *Modern Spatiotemporal Geostatistics* literature.

$$\boldsymbol{\chi}_{\text{map}} = Q(\boldsymbol{\chi}^0_{\text{map}}) \tag{4.18}$$

where $\chi_i = Q_i(\boldsymbol{\chi}^0_{\text{map}})$. The initial conditions are expressed as $\boldsymbol{\chi}^0_{\text{map}} = \boldsymbol{R}(\boldsymbol{\chi}_{\text{map}})$ or $\chi^0_i = R_i(\boldsymbol{\chi}_{\text{map}})$. The pdf model $f_{\mathcal{G}}$ is derived directly from the physical Eq. 4.18, as follows

$$f_{\mathcal{G}}(\boldsymbol{\chi}_{\text{map}}) = f_0(\boldsymbol{\chi}^0_{\text{map}})|J| \tag{4.19}$$

where $|J|$ is the corresponding Jacobian. In light of Eq. 4.19, one does not need to use assumption 4.3 or 4.16. ▪

4.2.4.3 *Non-Bayesian Conditionalization*

As we shall see in Chap. 5 and 6, there exist physical situations in *TGIS* practice in which a *non-Bayesian* conditionalization scheme may be worth attempting in *Step 4* of Sect. 4.2.1 above. This scheme could very well imply a few notable modifications of the original *BME* formalism. For instance, it turns out that certain conceptual issues may arise in some *TGIS* studies concerning the use of a Bayesian (evidence-based) conditionalization method vs. a non-Bayesian (truth functional-based) conditionalization method. In Fig. 5.8 of the subsequent Chap. 5 the interested reader will find a graphic display of a number of space/time mapping techniques which are developed within the context of the *Modern Spatiotemporal Geostatistics* paradigm.

4.2.5 Spatiotemporal Filtering

Yet another kind of an interesting modification which involves *Step 4* of the *BME* approach above is the so-called

BME filtering, in which soft data are available at the estimation points themselves.

A detailed theoretical analysis may be found in the relevant modern geostatistics literature. In this book we limit our presentation to an illustrative example, as follows.

Example 4.5. Numerical studies of particulate matter (PM_{10}) and ozone (O_3) distributions by Christakos and Serre (2000b) and Christakos et al. (2000) have shown that space/time filtering can lead to significant improvements in prediction accuracy. Assume that the specificatory *KB*, $\mathcal{S} = \mathcal{S}_0 \cup \mathcal{S}_1$, includes hard data as in Eq. 3.15 and probabilistic soft data of the form of Eq. 3.21, where $\mathcal{S}_0$: $\boldsymbol{\chi}_{\text{soft}} \in \boldsymbol{I}$ and $\mathcal{S}_1$: $\boldsymbol{\chi}_k \in I_k$. In other words, $\mathcal{S}_1$ implies that soft data are available at the estimation points themselves. Under these conditions, the posterior pdf is given by

$$f^{bc}_{\mathcal{K}}(\boldsymbol{\chi}_k) = A^{-1}\int_I d\boldsymbol{\chi}_{\text{soft}} f_{\mathcal{S}}(\boldsymbol{\chi}_{\text{soft}}, \boldsymbol{\chi}_k) e^{\mu_0 + \mu^T g} \tag{4.20}$$

where the normalization parameter is $A = \int_{I,Ik} d\boldsymbol{\chi}_{\text{soft}}\, d\boldsymbol{u}\, f_{\mathcal{S}}(\boldsymbol{\chi}_{\text{soft}}, \boldsymbol{u})\, f_{\mathcal{G}}(\boldsymbol{\chi}_{\text{data}}, \boldsymbol{u})$. There are various techniques for encoding soft data at the estimation points. Given hard PM_{10}

measurements, a technique (e.g., polynomial fitting or model simulation) can be used to derive PM_{10}-values at the estimation points. The new values, which are uncertain, can be used to develop soft data at these points (e.g., probability functions having these values as means). In order to provide a numerical illustration of the effect of soft PM_{10} data at the estimation points, Christakos and Serre (2000b) examined the following situation. Consider a set of points (monitoting stations) where PM_{10} values are available but considered unknown for the purposes of the analysis. These PM_{10} values were then estimated

a By assuming no soft data at the estimation points.
b By using probabilistic soft data at the estimation points (for illustration purposes, these probability data were generated so that they were consistent with the actual values at the estimation points).

The corresponding estimation errors, say $e^{(a)}$ and $e^{(b)}$, were calculated at each one of the estimation points for the cases (*a*) and (*b*), respectively. First, as should be expected, the results obtained at the monitoring stations (Table 4.4) show that the

Table 4.4. The effect of PM_{10} soft data at the estimation points on August 31, 1995. **a** Without soft data, **b** with soft data at the estimation points

Monitoring station	Estimation error (µg m^{-3})	
	a Without soft data	b With soft data
40	0.75	0.12
44	0.73	0.13
46	1.33	0.22
39	4.95	0.67
1	3.20	0.54

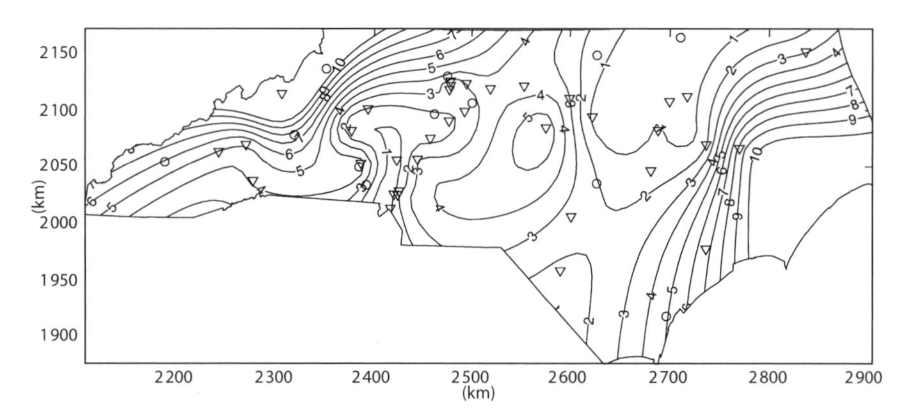

Fig. 4.6. Map of the Δe values (in µg m^{-3} of PM_{10}) over North Carolina averaged over a 3-day period (August 25, 31, and September 6, 1995). The map demonstrates the improvement gained by considering soft data at estimation points

approach (b) allows a better PM_{10} estimation accuracy than does the approach (a). Furthermore, in Fig. 4.6 we plot the distribution of the difference

$$\Delta e = e^{(a)} - e^{(b)} \tag{4.21}$$

in the estimation errors of the approaches (a) and (b) throughout the state of North Carolina averaged over a three-day time period. The reader may notice that the Δe values are consistently positive throughout the entire state (the Δe values vary from about 1.0 to 10.0 $\mu g\ m^{-3}$. Therefore, the Δe map strongly supports the argument that the incorporation of soft data at the estimation points (whenever available) can improve significantly the quality of PM_{10} estimation across space and time. ▪

4.2.6 Spatiotemporal Mapping and Change-of-Scale Procedures

In many cases, the adequate characterization of a life support field (natural, epidemiological, or social) may require an appreciation of its spatiotemporal variation at multiple scales. A rigorous description depends on the space/time scale at which the phenomenon is considered, whereas most fields only exist as meaningful entities over limited ranges of scale[11].

The scale of the investigation, which is observer-dependent, can play a major role in certain TGIS *applications.*

For illustrative purposes, Table 4.5 (from Christakos 2001) presents a situation in human exposure analysis, in which different scales have been used to classify fields, such as space, time, age, exposure level, health effects, and so on. Depending on the properties of the field under consideration (e.g., additive vs. non-additive fields)[12] and the application context considered (e.g., physical vs. epidemiological fields), the *change-of-scale* methods available to *TGIS* specialists include

- Straightforward methods (e.g., space/time averaging).
- Highly sophisticated methods (e.g., homogenization and renormalization).

Table 4.5. Scale variations considered in human exposure studies

Variable	Scale
Space	μm ⟶ km
Time	msec ⟶ years
Age	Young ⟶ old
Exposure	Micro ⟶ macro
Health effect	Local ⟶ global

[11] The appropriate emphasis on scale was given by Guye (1922) who made the acute remark that, "it is essentially the scale of observation that creates the phenomenon".

[12] Additivity implies that one can define the field values at the new scale as the average of the field values at the original scale.

These two classes of change-of-scale methods are examined below with the help of two examples. The first one (Example 4.6) is from epidemiology and the second one (Example 4.7) is from subsurface hydrology.

Example 4.6. In many epidemiologic studies, data at a higher scale (e.g., county level) are available, whereas the epidemiologist is interested in data at a lower scale (e.g., local zip code level) in which a more meaningful human exposure analysis may be performed. This situation requires some kind of

Down-scaling of TGIS*-based epidemiological modelling from the higher to the lower scale.*

In such cases, a useful down-scaling approach is as follows (see Fig. 4.7). Assume that the $X(V_i,t)$ represents the epidemiologic random field (e.g., the mortality rate) at the level of counties $V_i = V(\boldsymbol{s}_i)$, $i = 1, \ldots, N$ (each county is centered around a location specified by the vector $\boldsymbol{s}_i$). Let $X(\boldsymbol{u},t)$ represent the epidemiologic field at the local level (say, zip code) within V_i, i.e., $\boldsymbol{u} \in V_i$. In light of the additivity property, these two epidemiologic fields are related by,

$$X(V_i,t) = |V_i|^{-1} \int_{V_i} d\boldsymbol{u}\; X(\boldsymbol{u},t) \tag{4.22}$$

Assuming that the fields are spatially homogeneous and temporally stationary, it is easily shown that the corresponding field means are equal, i.e., $m_V = m_x$, where $m_V = \overline{X(V_i,t)}$ and $m_x = \overline{X(\boldsymbol{u},t)}$. By considering the field 4.22 at two different counties and time periods, and then taking the expected value of the product, we find that the corresponding field covariances are related as follows:

$$c_V(V_i,t;V_j,t') = \left[|V_i||V_j|\right]^{-1} \int_{V_i}\int_{V_j} d\boldsymbol{u}d\boldsymbol{u}'\, c_x(\boldsymbol{u},t;\boldsymbol{u}',t') \tag{4.23}$$

Since epidemiologic $X(V_i,t)$ data are available at the county V_i level, the m_V and c_V can be calculated. Then, $m_V = m_x$ and the c_x is obtained from Eq. 4.23 by direct inversion or by some other technique. An efficient technique is to assume several models for c_x and then calculate their parameters by fitting the models to Eq. 4.23; the c_x model that provides the best fit to the c_V data is selected. Given the $X(\boldsymbol{u},t)$ statistics, estimates of the epidemiologic field at the local zip code level can be obtained using the *BME* approach. For numerical illustration, let us assume that $X(V_i)$ data of a spatial death rate field are available at the county level (Fig. 4.8a). For exposure analysis purposes,

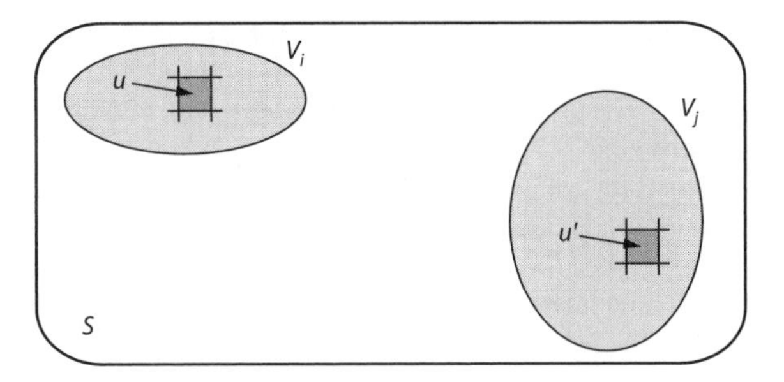

Fig. 4.7. An illustration of the down-scaling configuration for an epidemiological field. S = state, V_i, V_j = counties, and $\boldsymbol{u}$, $\boldsymbol{u}'$ = zip codes

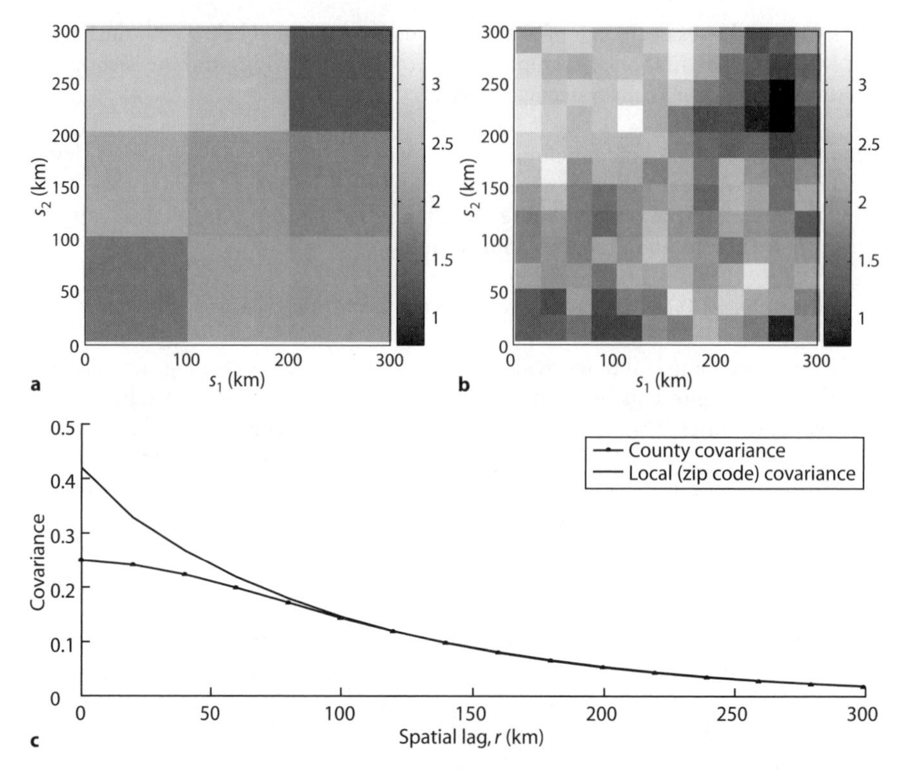

Fig. 4.8. Maps of death rates (deaths / day / 100 000 people) at **a** the county level; and **b** the local (zip code) level. **c** The corresponding covariance functions

death rate values $X(\boldsymbol{s})$ are often needed at the local (zip code) level. In this case, Eq. 4.22 reduces to the purely spatial expression

$$X(V_i) = \left|V_i\right|^{-1} \int_{V_i} d\boldsymbol{u} X(\boldsymbol{u}) \tag{4.24}$$

where the $X(V_i)$ mean is, $m_V = 2.0$ (deaths / day / 100 000 people), and its covariance is plotted in Fig. 4.8c. Using the down-scaling analysis outlined above we find that the mean value of $X(\boldsymbol{s})$ is, also, $m_x = 2.0$, and its covariance between any two spatial points $\boldsymbol{s}$ and $\boldsymbol{s}'$ is given by:

$$c_x(r) = c_{0,1} \exp(-3r / a_{r,1}) + c_{0,2} \exp(-3r / a_{r,2}) \tag{4.25}$$

where $r = |\boldsymbol{s} - \boldsymbol{s}'|$, $c_{0,1} = 0.4$ (deaths / day / 100 000 people)^2, $a_{r,1} = 300$ km, $c_{0,2} = 0.02$ (deaths / day / 100 000 people)^2, and $a_{r,2} = 20$ km. This covariance is also plotted in Fig. 4.8c, for comparison. Finally, the corresponding map of the local (zip code) death rates $X(\boldsymbol{s})$ is plotted in Fig. 4.8b. ▪

We now turn our attention to a different kind of a change-of-scale situation from subsurface hydrology.

Example 4.7. In the case of subsurface flow in an aquifer, heterogeneities occur at the scale of centimeters, while the scale of the entire aquifer is on the order of kilometers. Situations like this require that the *TGIS* uses very large numerical grids in space and time, which even the most advanced computers cannot handle, at present. Hence, the *TGIS* specialists need to use change-of-scale methods that permit the local scale heterogeneities to be processed faster. Some of these methods are briefly reviewed below (for a more detailed discussion, the interested reader is referred to the relevant literature):

- The *averaging* method uses smoothing and integration of the natural field over a specified space/time domain (e.g., a representative elementary domain) and can be used for natural fields which satisfy the additivity property[13]. If the original field refers to a microscale, the averaged field will refer to a macroscale (see e.g., Bloschl and Sivapalan 1995).
- The *homogenization* method provides the means to up-scale physical differential equations involving natural parameters that are non-additive (e.g., hydraulic conductivity). In this case, the problem of interest is embedded in a family of problems parametrized by a scale parameter, and then up-scaling is achieved by letting the scale parameter tend to zero (Bakhavlov and Pansenko 1989).
- The *renormalization group* (*RNG*) method accurately predicts large-scale properties by following the evolution of the pertinent quantities under change- of-scale transformations derived from physical laws. Renormalization makes it possible to investigate systematically the effects of local scale fluctuations on the large-scale properties of contaminant flow and transport (Hristopulos and Christakos 1999).

In Fig. 4.9, an example of the RNG concept is presented in the case of subsurface hydraulic conductivity K distribution which varies within a two-dimensional porous formation. The reasoning goes as follows. Assume that the microscale K values are available at sixteen small blocks. At each successive stage $n = 0$, 1 and 2, the previous K values are replaced by new K values within larger blocks whose area is the summation of the four previous blocks. Accordingly, the new K value is a function of the four previous ones and we can write

$$K = \Phi[K_1, K_2, K_3, K_4] \tag{4.26}$$

where the form of $\Phi[\cdot]$ depends on the specific RNG features assumed. The procedure is repeated several times until the final K_{eff}-value is found (in this example this happened at the stage $n = 2$). Both analytical and numerical RNG forms have been used in the relevant literature. In Fig. 4.10 the analytical RNG-based K_{eff} solution proposed by Hristopulos and Christakos (1999) is compared to the results by a widely used numerical technique. As is demonstrated in this figure, the RNG method provides a better fit to the measured data (in this case, the limestone and sandstone laboratory samples) than the popular numerical technique. ■

[13] I.e., it is physically meaningful to define the field values (e.g., amount of contaminant) at the new scale simply as the average of the field values at the original scale.

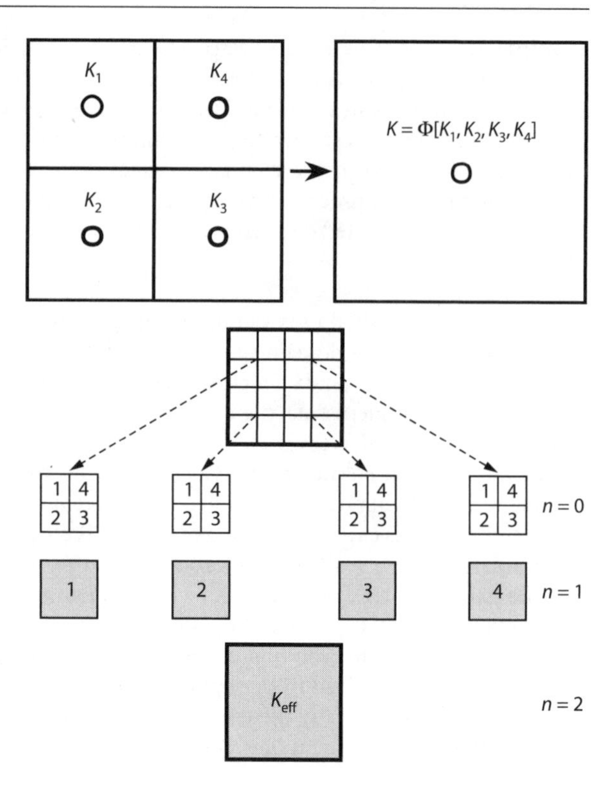

Fig. 4.9.
An illustration of the renormalization group (RNG) approach in the case of subsurface hydraulic conductivity

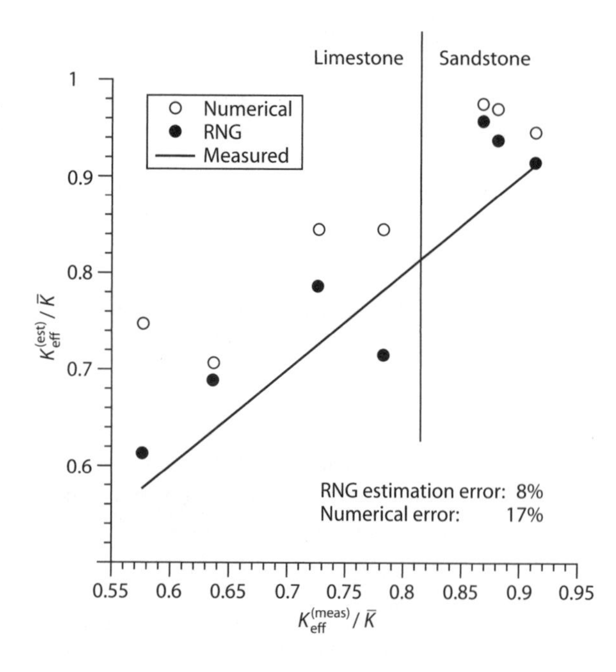

Fig. 4.10.
Comparison of K_{eff} obtained by RNG vs. that obtained by a numerical technique ($\bar{K}$ specifies the arithmetic mean of K)

A number of substantial change-of-scale issues arise naturally in several other scientific disciplines of *TGIS* interest. There exist biologic situations, e.g., in which the study of humans can only be performed over a certain range of coarse scales, whereas an organism looks completely different seen through a microscope, when individual cells become visible. In biodiversity investigations, in order to understand space/time variation at one specific scale the practitioner must study variations at several other scales (Mauer 1994). In ecosystem management, one needs to analyze a hierarchy of ecosystem scales (from national to site-specific) to permit the choice of detail that suits the management objectives and proposed use (see e.g., Bailey 1995). Finally, geomorphological studies of landscape involve the characterization of the surface form at several scales (Wood 1996).

4.3 Other Mapping Techniques

We briefly discuss some mapping techniques which are of interest to *TGIS* specialists. Then, a comparison is made of the main properties of each one of these technique with the relevant properties of the *BME* approach. The conditions are examined under which many of these techniques are derived as special cases of the *BME* approach.

4.3.1 Wiener-Kolmogorov Stochastic Interpolation

A group of space/time mapping techniques introduced by Petersen and Middleton (1965) traces back to the early *Wiener-Kolmogorov stochastic interpolation* theory (Kolmogorov 1939; Wiener 1949). The two main stages of stochastic interpolation are summarized below:

i. A linear estimator of the natural field in question is assumed, and the weights of the estimator are estimated so that the statistical mean-squared interpolation error is minimized.
ii. The weights are substituted back into the estimator and interpolation error equations, to calculate the interpolation values and their associated statistical accuracy.

Experimental data-based implementation of stochastic interpolation uses input parameters (covariance, variogram, etc.) calculated from the dataset available, whereas model-based implementation of stochastic interpolation usually involves the calculation of the input parameters from physical laws and scientific theories. The group of stochastic interpolation techniques includes:

a *Objective analysis* techniques of meteorological forecasting (Gandin 1963; Thiebaux and Pedder 1987);
b *Gauss-Markov* techniques that have been implemented extensively in oceanography (e.g., Bretherton et al. 1976);
c *Spatial Kriging* methods (e.g., Matheron 1965; Journel 1989), which have been routinely employed in mining, petroleum engineering and geohydrology, and to a lesser extent in *GIS*; and

d *Spatiotemporal estimation* techniques (Christakos 1992; Christakos and Hristopulos 1998), which have been used for space/time environmental monitoring and disease control purposes.

Due to their rather frequent use in *TGIS* applications, in the next section we will discuss some of the Kriging techniques (*c*) above. As it turns out, the Kriging techniques can be derived as special cases of the *BME* approach, a fact that testifies to *BME*'s considerable generalization power.

4.3.2 Geostatistical Kriging and Neural Networks

From the *TGIS* point of view, perhaps the most important function of classical geostatistics (Matheron 1965) is spatial and/or temporal estimation which is performed in terms of the so-called

Kriging techniques.

There exist various types of Kriging in the literature, including the following (the interested reader may consult the cited references for more details):

- *Ordinary* Kriging, (*OK*; Matheron 1965; Journel 1989). This is a widely used type of classical Kriging in cases where the spatial variation is spatially homogeneous and the mean function is not known or is difficult to estimate.
- *Simple* Kriging (*SK*; Matheron 1965; Journel 1989). It is used in place of *OK* when the mean function is known.
- *Intrinsic* Kriging (*InK*; Matheron 1973). An elegant technique that was originally developed to estimate nonhomogeneous spatial variations, although not extensively used in practice (perhaps due to the theoretical and computational challenges it presents).
- *Indicator* Kriging (*IK*; Journel 1983; Goovaerts 1997). This technique focuses on the estimation of indicator random fields, requires only local spatial homogeneity, and has been proven useful in the study of the spatial distribution of threshold functions.
- *Disjunctive* Kriging (*DK*; Rivoirard 1994). It is used to derive estimates relative to cut-off values. It has some similarities with *IK* (in fact, certain geostatisticians view *IK* as a special case of *DK*; e.g., Chiles and Delfiner 1999).

Despite their practical usefulness in certain applications, it is widely recognized nowadays that the techniques of classical geostatistics have a number of limitations which limit considerably their theoretical generality as well as their applicability in practice (for a review of these limitations, see Goovaerts 1997; Olea 1999; and Christakos 2000).

As it should be expected, the Kriging techniques are *special cases* of the considerably more general *BME* approach. In fact, some of the Kriging assumptions above determine the conditions under which a *BME* technique could be reduced to a specific Kriging technique. In Table 4.6 we depict situations in which the *BMEmode* esti-

Table 4.6. Comparing *BME* to some popular kriging estimators

$\mathcal{G}$	$\mathcal{S}$	*BME* estimate	
$\boldsymbol{\gamma}_{\text{map}}$	$\boldsymbol{\chi}_{\text{hard}}$	$\left.\begin{array}{l} \hat{\chi}_{k,\text{mode}} = \boldsymbol{\lambda}^T \boldsymbol{\chi}_{\text{hard}} = \hat{\chi}_{k,OK} \\ \lambda_i = \gamma_{ki}^{-1}\left[\sum_{j=1}^{m} \gamma_{kj}^{-1}\right]^{-1} \end{array}\right\}$	(4.27)
$\overline{\boldsymbol{x}}_{\text{map}}$, $\boldsymbol{c}_{\text{map}}$	$\boldsymbol{\chi}_{\text{hard}}$	$\left.\begin{array}{l} \hat{\chi}_{k,\text{mode}} = \overline{x}_k - \boldsymbol{\lambda}^T \boldsymbol{\chi}_{\text{hard}} = \hat{\chi}_{k,SK} \\ \lambda_i = -c_{ki}^{-1}\left[c_{kk}^{-1}\right]^{-1} \end{array}\right\}$	(4.28)
$\boldsymbol{\kappa}_{\text{map}}$	$\boldsymbol{\chi}_{\text{hard}}$	$\hat{\chi}_{k,\text{mode}} = \hat{\chi}_{k,InK}$	(4.29)

mate coincides with a Kriging estimate (a more detailed analysis and mathematical derivations can be found in Christakos 2000). In this table, the $\boldsymbol{\gamma}_{\text{map}}$, $\overline{\boldsymbol{x}}_{\text{map}}$, $\boldsymbol{c}_{\text{map}}$, and $\boldsymbol{\kappa}_{\text{map}}$ denote the matrices of space/time variograms, means, covariances, and generalized covariances, respectively; the $\hat{\chi}_{k,OK}$, $\hat{\chi}_{k,SK}$, and $\hat{\chi}_{k,InK}$ specify the *OK*, *SK*, and *InK* estimates, respectively; c_{ki}^{-1} and γ_{ki}^{-1} denote the ki^{th} element of the inverse matrix $\boldsymbol{c}_{\text{map}}$ and $\boldsymbol{\gamma}_{\text{map}}$, respectively; finally, λ_i are the corresponding estimation coefficients (note that *BME* provides explicit expressions for these coefficients)[14].

In light of the above results, one can conclude that by being a considerably more general approach, *BME* helped Kriging determine its domain of applicability. In other words,

BME does not discard Kriging but rather shows it to be only a special case of limited application within a more general concept.

Indeed, the improvement of ideas, rather than their outright rejection, is the scope of science-based *TGIS* analysis. Furthermore, the fact that Kriging models can be shown to be limiting cases of *BME* has some interesting consequences. One is *inertia*, i.e., many of the tools and tricks used by Kriging to estimate correlation functions, approximate spatial trends, etc. can be still useful in *BME* analysis. Inertia has a positive psychological effect, because *TGIS* scientists who invested a considerable amount of time and energy in learning geostatistical techniques will feel better knowing that they can still use them under the appropriate conditions outlined by the general *BME* theory. Another one is *evidence transfer*, i.e., in those specific cases that Kriging predictions are evidentially confirmed (to within a prediction error), the confirmation is transferred to *BME*. Finally, an important characteristic of modern *TGIS* analysis is its *continuity*, i.e., the ability of the new *BME*-based advanced functions to solve problems that could not be handled previously by the Kriging methods, thus providing convincing evidence of the ability of theoretical *TGIS* analysis to continuously generate powerful concepts and tools that improve our understanding of space/time phenomena.

[14]Other forms of kriging (probability and multigaussian kriging, co-kriging, etc.) can be also considered in the light of *BME*.

Like Kriging, neural networks are spatial analysis techniques which are concerned with data-driven learning for generalization purposes (Cherkassky and Mulier 1998). In particular, neural networks consist of several simple processing units (neurons) which may be trained (learning from data) and then used for prediction at unsampled points (generalization). Neural networks may be used to represent non-linear large scale trends and then Kriging is applied to the residuals (Kanevski et al. 1999). However, neural networks are based on a "black box" non-rigorous approach which suffers from a number of inherent limitations: model-free estimators, overparameterization, sensitivity to initial conditions and stopping rules, multiple local minima, misspecification of learning method, problems with the interpretability of the results, etc.

4.3.3 Kalman-Bucy Filtering

A useful group of space/time mapping techniques is based on the so-called *Kalman-Bucy* filtering concept (e.g., Kailath 1981). In a sense, these techniques can be seen as recursive formulations of the celebrated Wiener-Kolmogorov interpolation approach of Sect. 4.3.1 above. There are many variations of the Kalman-Bucy filtering. Roughly, the three main stages of the approach are as follows:

i. It is assumed that the law governing the natural process can be represented by a randomly perturbed multivariate first-order differential equation (dynamic system equation). The observation equation is assumed to be linear with respect to the natural process plus a random error term.
ii. The estimate at the current time is given as a weighted linear combination of previous estimates plus data at current time, and the weights are calculated so that the statistical mean-squared interpolation error is minimized.
iii. The weights are substituted back into the estimator and interpolation error equations, to obtain recursive forms of the estimator and its associated statistical accuracy.

The optimality of the Kalman-Bucy filter depends on the complete knowledge of a number of parameters (including the coefficients of the system and observation models, the statistics of the random perturbation and error terms, and the covariance and cross-covariance matrices). In most cases, the system and observation equations are assumed linear[15]. In some other applications the physics of the situation may lead to non-linear systems. Treatment of non-linear systems within the Kalman-Bucy filter formulation is an approximate approach, usually called the *extended* Kalman-Bucy filter (e.g., Bras and Rodriguez-Iturbe 1985). Extended Kalman-Bucy filters involve approximations in the form of piece-wise linearizations of the non-linear system and observation equations, and the underlying probability distributions are usually assumed to be of the Gaussian form. In the non-linear case, the Kalman-Bucy filter loses much of the elegance and comprehensiveness that it exhibited in the linear case.

When applied in a space/time domain certain issues arise regarding the *recursivity* of the Kalman-Bucy filter. Recursivity is usually maintained only with respect to time.

[15]In fact, the linear recursive Kalman-Bucy filter was originally developed for time series problems, where it was met with great success (e.g., Johnson and Dudgeon 1993).

Christakos and Bogaert (1996), e.g., assumed that the natural process $X(\boldsymbol{s},t)$ can be represented by the dynamic system equation

$$X(\boldsymbol{s}_k, \boldsymbol{t}_{\ell+1}) = \mathcal{P}_s[X(\boldsymbol{s}_k, \boldsymbol{t}_\ell)] + Z(\boldsymbol{s}_k, \boldsymbol{t}_\ell) \tag{4.30}$$

where $\mathcal{P}_s[\cdot]$ is a linear operator and $Z(\boldsymbol{s}_k,\boldsymbol{t}_\ell)$ is an *S/TRF* (e.g., white noise) with known statistics; $Z(\boldsymbol{s}_k,\boldsymbol{t}_\ell)$ is statistically independent of $X(\boldsymbol{s}_i,\boldsymbol{t}_j)$ for $j \leq \ell$. Then, the recursive equations of space/time prediction at $(\boldsymbol{s}_k,\boldsymbol{t}_{\ell+1})$ based on data at $(\boldsymbol{s}_i,\boldsymbol{t}_j)$, $i \neq k$, $j \leq \ell$, and the associated error variance are given by, respectively,

$$\hat{X}(\boldsymbol{s}_k, \boldsymbol{t}_{\ell+1}) = \mathcal{P}_s[\hat{X}(\boldsymbol{s}_k, \boldsymbol{t}_\ell)] \tag{4.31}$$

which offers a prediction of the future value of the process at time $\boldsymbol{t}_{\ell+1}$ in terms of the estimation at the present time $\boldsymbol{t}_\ell$, and

$$\sigma_x^2(\boldsymbol{s}_k, \boldsymbol{t}_{\ell+1}) = \mathcal{P}_s^2[\sigma_x^2(\boldsymbol{s}_k, \boldsymbol{t}_\ell)] + \sigma_z^2 \tag{4.32}$$

where σ_z^2 is the variance of Z.

Starting from the differential equation governing the distribution of sulfur dioxide, Omatu and Seinfeld (1981) used distributed modelling and space/time Kalman filter to predict air pollution. Christakos (1985) used vector Kalman filtering techniques in the prediction of soil profiles. Early attempts to apply these techniques in atmospheric data analysis include the works of Jones (1965) and Petersen (1973). During the last twenty years, space/time Kalman-Bucy filtering has been applied with considerable success in the study of meteorological problems. Multivariate space/time Kalman filtering has been used, e.g., by Ghil et al. (1981) in a longitude-time domain involving a midlatitude *f*-plane shallow water model linearized about a mean zonal flow, and by Parrish and Cohn (1985) in a latitude-longitude-time domain assuming a similar model on a midlatitude beta plane (a review of the meteorological applications of space/time Kalman filter can be found in Daley 1991).

4.3.4 Some Comparisons

Several geographical studies have provided useful comparisons of a number of interpolation and mapping techniques. Burroughs (1986), e.g., gives a summary of attributes and shortcomings of some deterministic and stochastic mapping techniques, including edge-seeking, Thiessen polygons, trend surface, Fourier series, splines, moving averages, and Kriging techniques[16]. *Modern Spatiotemporal Geostatistics* sought to revise the classical spatial interpolation paradigm in the light of epistemic concepts. One outcome of such a "revisionistic" effort was the *BME* approach. In this section we briefly compare the *BME* approach to other popular mapping techniques, and we point out some of its potential attractions in the context of *TGIS* modelling (for a comparative summary see also Table 4.7; additional details can be found in the relevant literature).

[16] Although, Burrough's discussion of these techniques refers to one-or two dimensional domains in space and does not concern composite space/time analysis.

Table 4.7. A comparison of mapping techniques (*OK*, *SK*, *InK*, *DK*, *IK* = ordinary, simple, intrinsic, disjunctive, and indicator Kriging, *OKB*, *EKB* = original, and extended Kalman-Bucy filters)

	BME	Kriging (OK, SK, InK, DK, IK)	Kalman-Bucy (OKB, EKB)
Estimation criterion	Epistemic	Statistical mean squared estimation error	Statistical mean squared estimation error
Estimator form	Arbitrary	– Linear (OK, SK, InK) – sum of univariate data functions (DK)	– Linear (OKB) – linearized approximation (EKB)
Physical law	Arbitrary form	Not considered	– Linear differential equation (OKB) – linearized (EKB)
Site-specific *KB*	– Hard data – observation model – intervals – probability functions – probabilistic logic – fuzzy sets	– Hard data – observation model – intervals (IK)	– Observation model
Statistical moments	Any order Multiple-point	1st and 2nd order Two-point	1st and 2nd order Two-point
Moment calculation	– Implicit in physical laws – explicit from physical laws	Explicit from experimental data	Explicit from experimental data or system dynamics
Underlying probability distribution	Any shape	– Gaussian – specific bivariate forms (DK, IK)	Gaussian
Space/time variation	– Homogeneous/ stationary – Nonhomogeneous/ nonstationary	– Homogeneous/ stationary (OK, SK, DK, IK) – Nonhomogeneous/ nonstationary (InK)	Homogeneous/stationary
KB conditionalization	– Bayesian – non-Bayesian	Linear statistics	Bayesian
Other estimation features	– Global – *KB* at estimation point		

In the remainder of this chapter, the term "classical Kriging" will serve as a summary term for some popular types of Kriging (i.e., the ordinary, simple, and intrinsic Kriging).

a *KB Integration*

The *BME* approach uses an epistemically sound criterion of knowledge integration and processing, whereas regression techniques (like Kriging, spatial statistics, time series, etc.) merely rely on a statistical criterion (e.g., the minimum mean squared error criterion). In its general form, *BME* can account rigorously and efficiently for several *KB* (both general and site-specific), which other techniques cannot consider or are limited to *ad hoc* solutions at best. In principle, the physical laws that can be considered by *BME* analysis are not restricted to a specified form. On the other hand, Kriging and other non-recursive type techniques do not directly incorporate physical laws[17]. Also, the Kalman-Bucy filter is a recursive technique that assumes physical constraints of a specific form (usually linear; Miller 1980)[18]. Some non-linear forms are approximated in terms of a linearization procedure, which usually involves a considerable loss of physical information. Classical types of Kriging and the Kalman-Bucy filter cannot account for soft data (of interval, probabilistic, etc. forms). The *IK* technique can incorporate certain interval types of data, although sometimes at the expense of involving certain restrictive approximations (Olea 1999).

b *Estimator Form*

The *BME* approach does not make any assumption about the form of the estimator. It can be linear or non-linear, in general, which implies that no restrictive assumptions are involved. The most commonly used classical Kriging and Kalman-Bucy techniques are limited to linear estimators. Certain non-linear estimator forms can be assumed in the case of *DK* and *IK* or in the case of the extended Kalman-Bucy filter and some neural network-based estimators, but this usually comes at the price of some kind of restrictive assumption on the form of the distribution, linearization approximations, overparameterization, interpretability problems, etc. (see Tanizaki 1993; Chiles and Delfiner 1999; Kanevski et al. 1999). The estimation error formulas of the classical Kriging and the Kalman-Bucy filter are data-independent, a property that has raised considerable criticism by many geostatisticians (see e.g., Goovaerts 1997).

c *Multi-Point Characterization*

The *BME* approach offers a complete characterization of the space/time map in terms of its multivariate pdf in the general case. As a result, it can consider multi-point mapping of several space/time points simultaneously for a variety of soft data combinations and probability distributions of arbitrary shapes. In a recent article titled "Generating probabilities in support of societal decision-making", Pfaff and Peteet made the following suggestion (Pfaff and Peteet 2001):

> We see a role also for changing the information that scientists feed to decision makers. In particular, scientists could better serve society by generating not only possible scenarios, but also probabilities, even for decisions that are to be made in the near future.

This is, in fact, what the *BME* approach can provide in terms of the multivariate pdf associated with the possible space/time maps. In general, neither classical Kriging nor Kalman-Bucy filtering can accomplish such a characterization (e.g., certain Kriging estimates could be obtained at multiple points in the special case of a multigaussian distribution, although this case is not commonly encountered in practice). Furthermore, the *BME* approach can account for soft data at the estimation points, a feature that usually improves the mapping accuracy significantly.

---

[17] In some cases the variogram can be calculated from a physical law – although this is not a common practice in traditional Geostatistics; Kitanidis (1997).

[18] E.g., multivariate first-order differential equations.

d *Probability Distribution Shape*

The *BME* approach does not make any assumption regarding the shape of the underlying probability distributions, which means that non-Gaussian multivariate distributions are automatically included in the *TGIS* analysis. On the other hand, underlying the classical Kriging and the Kalman-Bucy filter is the assumption of a Gaussian probability distribution. The *DK* and *IK* can incorporate certain forms of non-Gaussian (isofactorial) bivariate distributions, although some difficulties arise when a suitable isofactorial model needs to be constructed in practice (Chiles and Delfiner 1999). While *IK* does not require prior modelling of the probability distribution, a considerable loss of information is inevitable by replacing the original data by indicator data. Also, unfeasible values for the probability distribution may be obtained by *IK*, and the large number of thresholds involved require a heavy computational effort (see also Sect. 6.5.2).

e *Spatiotemporal Heterogeneity Patterns*

The *BME* approach can handle all forms of space/time heterogeneity, including spatially homogeneous or non-homogeneous trends, and stationary or non-stationary patterns. Most types of Kriging techniques (including *OK*, *SK* and *DK*) and the Kalman-Bucy filter do not apply directly to non-homogeneous/non-stationary patterns. The space/time trends must be determined and subtracted from the original pattern, before these Kriging and Kalman-Bucy techniques can be used. In many physical applications, trend determination is a process which often creates significant problems at both the conceptual and the computational levels (see e.g., Matheron 1965). The rigorous theoretical background of *InK*, on the other hand, allows it to handle successfully rather complicated spatially nonhomogeneous patterns. Unfortunately, the advanced mathematical theory of *InK* is considered by many to be the main reason for its rather limited popularity among practicing geostatisticians. Moreover, *BME* can incorporate multiple-point statistics (linear and non-linear) and higher-order moments in space/time in a rigorous and physically meaningful manner, whereas most stochastic interpolation techniques are limited to two-point statistics. Also, while in some pattern recognition applications neural networks may account for certain forms of multiple-point statistics, their implementation usually requires training images which are unrealistically similar to the actual (but unknown) image, etc. (see also discussion in Sect. 4.3.2 above).

Table 4.8.
Comparisons of computational efforts for the various mapping techniques. (*OK*, *SK*, *InK*, *DK*, *IK* = ordinary, simple, intrinsic, disjunctive, and indicator Kriging, *OKB*, *EKB* = original, and extended Kalman-Bucy filters)

Method	Computational effort
OK, SK	Reasonable
InK	Reasonable to heavy
DK, IK	Heavy to extremely heavy
OKB	Reasonable to very heavy
EKB	Heavy to extremely heavy
BME	Reasonable to very heavy

f *Computational Issues*

The assessment of the computational work required by each one of the techniques shown in Table 4.8 was based on the authors' own experience as well as on the collective experience of several other experts (e.g., discussions in Burrough 1986, and Daley 1991; among others). From a computational viewpoint, the *BME* approach generally requires the same effort as the classical Kriging techniques, when the same types of *KB* are considered. Naturally, a heavier computational effort is demanded when several additional types of soft data and general *KB* are processed by the *BME* approach (which, of course, leads to improved estimates across space and time). Fewer parameters are involved in *BME* than in the Kalman-Bucy filter, which naturally implies a lower computational effort. The computational implementation of the Kalman-Bucy filter presupposes that all parameters of the model and the filter are known. Incorrect specification of some of these parameters (e.g., the statistics of the system noise and the state random term) commonly leads to suboptimality and inconsistent filter behavior (e.g., divergence; see Bras and Rodriguez-Iturbe 1985). In realistic space/time domain applications, the Kalman-Bucy formulation is known to be computationally expensive even when efficiently formulated (the prediction error covariance matrix involves the multiplication of extremely large matrices in four-dimensional domains, etc.). Things get worst in the non-linear case. At this point, we would like to remind the reader that a number of examples providing numerical comparisons between *BME* and other space/time mapping techniques are discussed in detail in Chap. 6 ("The *BME* Toolbox in Action"). The reader can reproduce the results of these examples with the help of the *BMElib* software included for this purpose (see Chap. 7). To be sure, not all variants of computational *BME* can be possibly accommodated in one chapter, but our hope is that the readers will obtain a reasonably good idea of the various possibilities available to them.

4.4 Concluding Remarks

In describing his views on scientific development, Richard Phillips Feynman suggested that,

> One has to have the imagination to think of something that has never been seen before, never been heard of before. At the same time the thoughts are restricted in a straitjacket, so to speak, limited by the conditions that come from our knowledge of the way nature really is. The problem of creating something which is new, but which is consistent with everything which has been seen before, is one of extreme difficulty.

Feynman's insightful remark brings to the fore an important element of scientific development: The growth of knowledge consists of having ever more general techniques that produce the successful results of the previous techniques under similar conditions and they, also, account for ever more situations which the previous techniques could not account for. This is the case, indeed, of the *BME*-based techniques of *TGIS*. In light of the developments described in the previous sections, we would like to conclude the present chapter by noticing that

Among the main attractions of the BME *approach is its generality which surpasses that of any other existing space/time mapping method.*

(The existing mapping methods include the ones summarized in Table 4.7, as well as "black box" methods like basis functions and neural networks.

To put it in another way, the *BME* approach has enlarged considerably the physical domain in which the stochastic mapping techniques apply. This is, perhaps, one of the most substantial contributions of the *BME*-based advanced functions to modern *TGIS* modelling and mapping.

Chapter 5

Interpretive *BME*

Research is to see what everybody has seen,
and to think what nobody else has thought.
A. Szent-Gyorgyi

5.1 Interpretive Issues

In the *formal* analysis of the *BME* approach discussed in the previous chapters, certain assumptions expressed in a formal language, were laid down; precise rules of *KB* processing were formulated; pdf models were derived from the assumptions by means of these *KB* processing rules; and, finally, estimates were derived by means of certain optimization principles. No particular empirical meaning was given to the mathematical terms involved in formal analysis and no interpretation was attempted for the mathematical assumptions, such as Eq. 4.3. In other words, formal analysis is concerned with questions like "Does *A* follows from *B*?" rather than with questions of the form "Is this representative of the real world situation?"

Interpretive BME, on the other hand, is concerned with salient issues of physical meaning and interpretation[1]. Interpretation issues are relevant, e.g., when the *TGIS* user needs to

- Establish *correspondence* relationships between the life support systems and the formal mathematics which describe them.
- Measure and test the formal structure.
- Define a relationship between the observer and the observed.
- Justify certain methodological steps of the *TGIS* mapping procedure.

Other aspects of the interpretive *BME* approach – in which the formal steps receive a physical interpretation – will be also examined in the present as well as in the following chapters, occasionally with the help of some real world applications. As a matter of fact, the interpretive *BME* characteristics constitute one of the major paradigmatic differences between *Modern Spatiotemporal Geostatistics* and classical spatial statistics (see also discussion in Sect. 1.4)[2].

[1] Generally speaking, in formal analysis a conclusion is a matter of proof, whereas in interpretive analysis a conclusion is a matter of good physical evidence and an effective cognitive scheme.

[2] Of course, the proper interpretation of the *BME* approach is not merely a philosophical issue, but it has significant practical consequences as well. In some cases, the unfortunate impression has been created that the entropic concept is used by analogy with thermodynamics, while it is actually an information-theoretic or an epistemic probability concept. Apropos, philosophical investigations should be based on a sound understanding of the physics of the situation, otherwise they may lead to inappropriate conclusions.

5.2
An Epistemic Analysis of the *BME* Approach

The scientific method explicitly recognizes the power of *epistemology* to study the structure of knowledge, to obtain answers to questions related to the acquisition, meaning, and processing of *KB*, and to use mathematical abstractions to represent complex natural systems. A passage from the recent book of Kitchin and Tate (2000) titled "Research in Human Geography" is appropriate here:

> To make things more complex, research is rarely just a process of generating data, analysing and interpreting the results. By putting forward answers to research questions you are engaging in the process of debate about what can be known and how things are known. As such, you are engaging with *philosophy*.

In such a context, the justification of certain mathematical abstractions and assumptions made by *BME* may involve the use of epistemic tools (a detailed account is given in Christakos 2000). More specifically, an epistemic justification of the mathematical form of $f_{\mathcal{G}}$ assumed in Eq. 4.3 is obtained by embracing the following requirement:

R1: A maximally informative model $f_{\mathcal{G}}$ *given the general KB* ($\mathcal{G}$).

The above requirement involves the concept of "information". If the information measure chosen is based on the *Shannon* information concept (properly modified to account for the space/time mapping environment), the informativeness requirement is mathematically expressed in terms of a *maximum entropy* type of condition. The latter condition is the reason for the "*ME*"-part in the acronym *BME*, suggested by Christakos (1990). This is, however, one of the options allowed by the *Modern Spatiotemporal Geostatistics* paradigm. If a different information measure is used, the resulting mathematical formulation of the requirement *R1* above may not necessarily involve an entropy-type function. For example, in Sect. 4.2.4 we discussed the case in which the Shannon information is replaced by the *Fisher* information concept. Then, the requirement *R1* generally leads to different probability models $f_{\mathcal{G}}$, although the same results are obtained in certain remarkable cases (see e.g., Example 4.3). In comparing the current *BME* formulation of *R1* to other options, it must be noted that *BME* has certain important virtues: it leads to a simpler yet powerful set of equations, has appealing analytical and computational features, and is epistemically well-grounded[3].

Comment 5.1. The reader may find it interesting that the requirement *R1* can be also interpreted in terms of the *action principle* approach. In traditional language, an action principle seeks a criterion by means of which the result of a process is achieved (Lemons 1997). Many important laws of nature are derived by postulating an action principle. Most of modern physics, indeed, can be written in terms of principles, which are mathematical expressions of a certain kind of action (including

[3] We will not pursue any further the topics of epistemic analysis and information theory here. Instead, the reader is referred to the existing *Modern Spatiotemporal Geostatistics* literature.

Fermat's principle of least time, Jacobi's principle of least action, and Hamilton's principle). In *BME* analysis, the action is expressed in terms of the expected information given the $\mathcal{G}$-*KB* and the result (pdf model) is determined by maximizing this action principle.

Several years ago, the British mathematician and philosopher Alfred North Whitehead wisely remarked that,

> We think in generalities, and we live in detail.

Mutatis mutandis, the second requirement of interpretive *BME* analysis is concerned with model conditionalization in the light of site-specific knowledge, as follows:

R2: An updated model $f_{\mathcal{K}}^{bc}$ *conditioned on the site-specific KB* ($\mathcal{S}$).

In Christakos (1990, 1992) the $\mathcal{S}$-conditionalization of requirement *R2* is carried out by means of the epistemic assumption of *Bayesian* conditionalization, which explains the "*B*"-part in the acronym *BME*. In the context of *TGIS* advanced functions, the concept of $\mathcal{S}$-conditionalization is used to express the probability of the *S/TRF* event

$$X_k = \{x_k \leq \chi_k\} \tag{5.1}$$

assuming that another relevant event

$$X_{\text{data}} = \{x_{\text{data}} : \chi_{\text{data}} \in \mathcal{S}\} \tag{5.2}$$

has occurred ($\mathcal{S}$ refers to the site-specific *KB* described in Sect. 3.3.1). The assumption of Bayesian conditionalization (*bc*) implies that

$$P_{\mathcal{K}}^{bc}[x_k \leq \chi_k] = P_{\mathcal{G}}[X_k | X_{\text{data}}] \tag{5.3}$$

where

$$P_{\mathcal{G}}[X_k | X_{\text{data}}] = P_{\mathcal{G}}[X_{\text{data}} \wedge X_k] / P_{\mathcal{G}}[X_{\text{data}}] \tag{5.4}$$

The meaning of definition 5.3–5.4 is that the unconditional probability $P_{\mathcal{K}}^{bc}$ of the event X_k is equal to the conditional probability $P_{\mathcal{G}}$ of the same event given X_{data}. After some mathematical manipulations, definition 5.3–5.4 leads to Eq. 4.4. The reader should be aware that the applicability of definition 5.3–5.4 in the physical situation under consideration should be verified, either (*i*) by comparing it to another way of evaluating the conditional probability, or (*ii*) by examining its consistency with physical and logical arguments that apply in the situation. Thus, a natural consequence of item (*i*) above is the key question:

How can the *bc* definition 5.3–5.4 be verified by means of a physically and logically meaningful conditional probability that is independently calculated?

We will demonstrate the significance of this question with the help of two examples. The first is a simple card game in which the evidence has a purely statistical nature.

Example 5.1. In the elementary game of a shuffled pack of 52 cards consider two events: "The first card drawn from the pack is an ace" or $X_{\text{data}} = \{x_1 = 1\}$, and "The second card will be an ace too" or $X_k = \{x_2 = 1\}$. Since there are 51 cards left after the first event, among which 3 are aces, it is clear that the required probability for the second event is 3/51. This is the result that we want to compare with definition 5.3–5.4. According to Eq. 5.4,

$$P_{\mathcal{G}}\left[X_k \middle| X_{\text{data}}\right] = \frac{4}{52} \times \frac{3}{51} / \frac{4}{52} = \frac{3}{51}$$

where the denominator and numerator have been calculated using elementary combinatorial formulas. Thus, in this case the *bc* is verified by an independent logical procedure. ▪

The following example serves to illustrate a class of physical applications where a belief-based Bayes analysis may not be useful in the study of uncertainty.

Example 5.2. Let x be a random variable representing the total amount of a fluid in a hydraulic system and assume that x is uniformly distributed between 0 and 9 (in liters). An experiment is performed where the hydraulic system is emptied by means of two siphons that collect the fluid in two separate vessels. Let x_1 and x_2 be the amounts of fluid, respectively, in the first and the second vessel at the end of the experiment. Then, $x = x_1 + x_2$. One of the siphons has twice the section of the other which, according to Bernoulli's law, implies that 1/3 of the fluid out of the hydraulic system goes to one of vessel and 2/3 of the fluid to the other vessel, but we do not know which of the siphons is connected to which vessel. This is a case in which the uncertainty is introduced by the incomplete knowledge about the experimental setup. Indeed, according to the experiment, we have two equally probable cases: $x_2 = 2x_1$ (in which case $x = 3x_1$), and $x_1 = 2x_2$ (in which case $x = (3/2)x_1$). If $X_{\text{data}} = \{x_1 < 2\}$ and $X_k = \{x < 2 \text{ or } x > 7\}$, then

$$P_{\mathcal{G}}\left[X_k\right] = \frac{4}{9}$$

and

$$P_{\mathcal{G}}\left[X_{\text{data}}\right] = \frac{1}{2} P_{\mathcal{G}}\left[\frac{1}{3}x < 2\right] + \frac{1}{2} P_{\mathcal{G}}\left[\frac{2}{3}x < 2\right] = \frac{1}{2}\left(\frac{6}{9} + \frac{3}{9}\right) = \frac{1}{2}$$

Furthermore, it can be shown (Aerts 1985) that the conditional probability of X_k on X_{data} calculated in a physically meaningful way is equal to 1/2. This probability value is different than that obtained from Eq. 5.4, which is

$$P_{\mathcal{G}}\left[X_k \middle| X_{\text{data}}\right] = P_{\mathcal{G}}\left[x < 2\right] / \frac{1}{2} = \frac{4}{9}$$

Hence, the Bayesian analysis does not apply in this case. ▪

Note that in both the above two examples the verification of the definition established by Eqs. 5.3–5.4 would be futile if we did not have another way to calculate the conditional probability. Certainly, there exist situations where it is not possible to find another way to calculate the conditional probability, in which case 5.3–5.4 remains an assumption, until its validity is established by means of the meaningful conclusions and useful results it leads to.

Definition of Eq. 5.3–5.4 above suggests *one possible* way to express conditionalization of the random field event X_k on the event X_{data}. This definition is based on statistical reasoning and is not generally connected with physical causality (see e.g., discussion in Glymour 1981). The X_{data} and X_k, on the other hand, belong to the same natural field $X(p)$ and, hence, they can be considered causally connected through a physical law. This being the case, the problem is that the evidential character of bc at the $\mathcal{S}$-based stage may be not able to express such a law in a physically consistent manner. This discussion brings to the foreground a particularly appealing feature of *BME*:

The BME approach offers a substantial improvement – compared to classical Bayesian conditionalization methods – by making sure that a physical connection has been taken into consideration at the $\mathcal{G}$-KB stage.

Indeed, the reader may recall that the $\mathcal{G}$-based stage leads to the formulation of the probability function $P_{\mathcal{G}}$, which is subsequently used in Eq. 5.3[4].

Modern Spatiotemporal Geostatistics establishes a sound framework of considerable flexibility. In certain cases, bc may be not the only choice available to the *TGIS* specialist. As is pointed out in Christakos (2000), other possibilities exist which involve the assumptions of material *S/TRF* conditionalization, biconditionalization, physical conditionalization, etc., thus leading to a *non-Bayesian* approach of *KB*-conditionalization. A brief summary of these alternative conditionalization assumptions is given in the following section. However, it is not our goal to provide a detailed discussion of these important issues in this volume.

5.3 Non-Bayesian Conditionalization

Since the definition of conditional probability ought to be consistent with the physical situation it describes, it should not come as a surprise that there is no unique definition of conditionalization that applies to all possible situations. Indeed, $\mathcal{S}$-conditionalization does not need to be restricted to the Bayesian assumption. It may involve alternative assumptions, thus leading to a *non-Bayesian* approach of *KB*-conditionalization. The text that follows briefly discusses these alternatives. In particular, a group of non-Bayesian conditionalization methods can be developed in terms of truth functionals (e.g., the X_{data} and X_k are related in terms of the truth functional connectives of negation, conjunction and disjunction), which establish a kind of a causal relevance for the *S/TRF* in question.

Comment 5.2. An important feature of truth functional-based *S/TRF* conditionalization is that it is a necessary condition for the validity of many other conditionalization schemes, including non-truth functional ones (e.g., physical, logical, and statistical conditionals).

[4] Hence, in many cases *BME* may not need to consider any further physical connection at the $\mathcal{S}$-based stage.

5.3.1 Material Biconditionalization

For illustration, in this section we will first analyze one of the most important truth functional approaches, namely, the *material biconditionalization* (*mb*; denoted by $\longleftrightarrow$). Also, later in the section we will describe the *material conditionalization* (*mc*). The *mb* is based on a truth functional of the form $X_{data} \longleftrightarrow X_k$, meaning that "the *S/TRF* events X_{data} and X_k are either both true or both false". The *mb* probability of the map is defined by

$$P_{\mathcal{K}}^{mb}[x_k \leq \chi_k] = P_{\mathcal{G}}[X_{data} \longleftrightarrow X_k] \tag{5.5}$$

where

$$P_{\mathcal{G}}[X_{data} \longleftrightarrow X_k] = 2P_{\mathcal{G}}[X_{data}, X_k] + 1 - P_{\mathcal{G}}[X_{data}] - P_{\mathcal{G}}[X_k] \tag{5.6}$$

Equation 5.5 is notably different from Eq. 5.3, which is the result of the fact that while the former presupposes some dependence (physical or otherwise) between the events X_{data} and X_k, the latter does not necessarily involve any such dependence. This difference can turn out to be crucial in some physical situations. As is shown below, in the case of the hydraulic system described in Example 5.2, the *mb* provides the physically correct result (which was not the case with *bc*).

Example 5.3. In the case considered in Example 5.2 above, the *mb* of Eq. 5.6 gives,

$$P_{\mathcal{G}}[X_{data} \longleftrightarrow X_k] = 2\left(\frac{4}{9} \times \frac{1}{2}\right) + 1 - \frac{1}{2} - \frac{4}{9} = \frac{1}{2}$$

which coincides with the correct conditionalization result for the hydraulic system. ▪

If *mb* is used at the posterior stage of the space/time mapping process, the resulting modern geostatistical techniques include:

- The *MbME* technique, i.e., *Material biconditionalization Maximum Entropy*, which is considered in this chapter.
- The *MbMF* technique, i.e., *Material biconditionalization Maximum Fisherian* (epistemic sense) technique. The latter feature is equivalent to the classical minimum Fisher information rule in the statistical sense.

The *mb* probability can be also expressed in terms of the statistical moments of the indicator *S/TRF*, which may offer an efficient means to estimate this probability in practice.

Example 5.4. Let us consider the scalar case, $X_{data} = X_1 \leq \chi_1$ and $X_k \leq \chi_k$. Then, Eq. 5.6 can be written as,

$$P_{\mathcal{K}}^{mb}[x_k \leq \chi_k] = 1 - \overline{I(X_1)} - \overline{I(X_k)} + 2\overline{I(X_1)I(X_k)} = 1 - 2\gamma_I(p_1, p_k) \tag{5.7}$$

where

$$\gamma_I(p_1, p_k) = \frac{1}{2}\overline{[I(X_1) - I(X_k)]^2}$$

is the variogram of the indicator field I between the points $\boldsymbol{p}_1$ and $\boldsymbol{p}_k$. Note that, $I(X_i) = 1$ if $X_i \leq \chi_i$ (at $\boldsymbol{p}_i$; $i = 1,k$); = 0 otherwise. ■

For *S/TRF* estimation purposes we need to invent an *mb*-based probability density for the posterior $\mathcal{K}$-stage. Assuming that $X_{\text{data}} = (\boldsymbol{\chi}_{\text{data}}, \boldsymbol{\chi}_{\text{soft}})$, the pdf associated with Eq. 5.6 is defined as follows

$$f_{\mathcal{K}}^{mb}(\boldsymbol{\chi}_k) = (2A - 1)^{-1}\left\{2Af_{\mathcal{K}}^{bc}(\boldsymbol{\chi}_k) - f_{\mathcal{G}}(\boldsymbol{\chi}_k)\right\} \tag{5.8}$$

where $A = P_{\mathcal{G}}[X_{\text{data}}]$ is the normalization parameter. The subscript $\mathcal{K}$ in Eq. 5.8 denotes that the general as well as the site-specific knowledge have been taken into account in defining the functional form of $f_{\mathcal{K}}^{mb}$. Equation 5.8 differs from the *bc* Eq. 4.4. A more detailed form of Eq. 5.8 is described in the following example.

Example 5.5. A particularly tractable form of $f_{\mathcal{K}}^{mb}$ is obtained if we assume only soft data, $X_{\text{data}} = \boldsymbol{\chi}_{\text{soft}}$ (something that is often the case in practice). Then, Eq. 5.8 reduces to the useful expression

$$f_{\mathcal{K}}^{mb}(\boldsymbol{\chi}_k) = (2A - 1)^{-1}\left[2\int_I d\Xi_{\mathcal{S}}(\boldsymbol{\chi}_{\text{soft}}) f_{\mathcal{G}}(\boldsymbol{\chi}_{\text{soft}}, \boldsymbol{\chi}_k) - f_{\mathcal{G}}(\boldsymbol{\chi}_k)\right] \tag{5.9}$$

where $A = \int_I d\Xi_{\mathcal{S}}(\boldsymbol{\chi}_{\text{soft}}) f_{\mathcal{G}}(\boldsymbol{\chi}_{\text{soft}})$. From Eq. 5.9 we find that, in the special case in which $f_{\mathcal{G}}(\boldsymbol{\chi}_{\text{soft}}, \boldsymbol{\chi}_k) = f_{\mathcal{G}}(\boldsymbol{\chi}_{\text{soft}}) f_{\mathcal{G}}(\boldsymbol{\chi}_k)$, we get $f_{\mathcal{K}}^{mb}(\boldsymbol{\chi}_k) = f_{\mathcal{K}}^{bc}(\boldsymbol{\chi}_k) = f_{\mathcal{G}}(\boldsymbol{\chi}_k)$. ■

Due to its truth functional-based definition, the $f_{\mathcal{K}}^{mb}$ is not a pdf in the "conventional" sense (as discussed, e.g., in Chung 1968). Indeed, while $P_{\mathcal{K}}^{mb} \in [0,1]$ and the choice of the normalization constant B is such that $\int_{-\infty}^{\infty} d\boldsymbol{v} f_{\mathcal{K}}^{mb}(\boldsymbol{v}) = 1$, it is nevertheless possible that $f_{\mathcal{K}}^{mb} < 0$ for some $\boldsymbol{v}$ values (e.g., if the $2A - 1$ takes negative values). The function $f_{\mathcal{K}}^{mb}$ may be viewed as a vehicle for calculating $P_{\mathcal{K}}^{mb}$, as well as for obtaining space/time estimates (in terms, e.g., of its mode, mean, and median values; see Example 5.8 below). In light of the above considerations, the $f_{\mathcal{K}}^{mb}$ would be characterized as a *pseudo-pdf* rather than as a pdf in the "conventional" sense[5]. Some analytical comparisons between *mb* and classical probability are made in the following example.

Example 5.6. Let us assume that the site-specific X_{data} reduces to $\boldsymbol{x}_{\text{data}} \leq \boldsymbol{\chi}_{\text{data}}$. Then, the *mb* probability and the *mb* pseudo-pdf are linked in a manner of considerable physical importance as follows

$$P_{\mathcal{K}}^{mb}\left[x_k \leq \boldsymbol{\chi}_k\right] = (1 - A) + (2A - 1)\int_{-\infty}^{\boldsymbol{\chi}_k} d\boldsymbol{v} f_{\mathcal{K}}^{mb}(\boldsymbol{v}) \tag{5.10}$$

where the coefficient A expresses the probabilistic structure of $\boldsymbol{\chi}_{\text{data}}$. Compare Eq. 5.10 with the classical *mb* probability

$$P_{\mathcal{K}}^{bc}\left[x_k \leq \boldsymbol{\chi}_k\right] = \int_{-\infty}^{\boldsymbol{\chi}_k} d\boldsymbol{v} f_{\mathcal{K}}^{mb}(\boldsymbol{v})$$

[5] In fact, "non-conventional" probabilities are not foreign to physical scientists (see e.g., the negative probability functions considered in Khrennikov 1997).

In the limited case that $x_{\text{data}} \leq \infty$, we get $A = 1$ and $P_{\mathcal{K}}^{mb}[x_k \leq \chi_k] = F_{\mathcal{K}}^{mb}(\chi_k) = F_{\mathcal{G}}(\chi_k)$, which is also the *bc* cdf. This should be expected, since by assuming all possible values for χ_{data}, we eliminate the consideration of any influence (physical, etc.) that the *KB*-specified values of χ_{data} may have on χ_k. In other words, conditionalization on site-specific data of the form $x_{\text{data}} \leq \infty$ naturally has no effect on the shape of the prior probability function. ▪

Since the *bc* pdf and the *mb* pseudo-pdf have different mathematical forms, in general, this implies that considerably different *S/TRF* estimates can be derived from these pdf (in terms of their mode, mean, etc.).

Example 5.7. Consider, for simplicity's sake, the single-point (χ_k) case in which the *mb* pseudo-pdf is a unimodal non-negative function. The *S/TRF* estimate $\hat{\chi}_k$ of interest at a point $\boldsymbol{p}_k$ can be defined as the solution of the equation

$$df_{\mathcal{K}}^{mb}(\chi_k)\,/\,d\chi_k = (2A-1)^{-1}\left[2A\,df_{\mathcal{K}}^{bc}(\chi_k)\,/\,d\chi_k - df_{\mathcal{G}}(\chi_k)\,/\,d\chi_k\right] = 0 \tag{5.11}$$

at $\chi_k = \hat{\chi}_{k,\text{mode}}^{mb}$[6], whereas for the *bc*-based pdf, the corresponding mode estimate $\hat{\chi}_k$ is the solution of the equation

$$df_{\mathcal{K}}^{bc}(\chi_k)\,/\,d\chi_k = 0 \tag{5.12}$$

at $\chi_k = \hat{\chi}_{k,\text{mode}}^{bc}$. Clearly, $\hat{\chi}_{k,\text{mode}}^{mb} \neq \hat{\chi}_{k,\text{mode}}^{bc}$. Intuitively, the evidential relevance expressed by *bc* is different than the causal relevance expressed by *mb*, in the sense that purely statistical conditionalization is not the same as space/time causation. ▪

In the following example, plots of the *bc* and *mb* probability densities are generated using data from the Chernobyl fallout study by Christakos et al. (2000).

Example 5.8. Radioactive soil contamination remains one of major impacts of the Chernobyl accident (Ukraine) on the environment and on human health. Contamination by the most widespread radionuclide, the Caesium 137 (^{137}Cs), covers large areas of Europe. The study by Christakos et al. (2000) focused on the Bryansk region (Fig. 5.1), which is the most affected territory in Russia. An extensive monitoring network was used (including 1745 sampling locations throughout Bryansk; see Fig. 5.2). Among other things, this network produced several probabilistic (soft) data of the mathematical form described in Table 3.1. In this study we focused on the distribution of the average ^{137}Cs concentration values for the year 1999. These values were calculated for a set of subregions in the Bryansk region. Each one of the subregions, say $V(\boldsymbol{s})$, was typically characterized by the spatial coordinates of its center-point $\boldsymbol{s} = (s_1, s_2)$. Then, the average ^{137}Cs concentration for $V(\boldsymbol{s})$ during the year $t = 1999$ was calculated as follows

$$C_V(\boldsymbol{s},1999) = |V|^{-1}\int_{V(\boldsymbol{s})} d\boldsymbol{u}\,C(\boldsymbol{u},1999) \tag{5.13}$$

[6] In some cases, the *mb* pseudo-pdf has a multimodal shape. Then, Eq. 5.11 may lead to more than one solutions, which may require a more sophisticated mathematical analysis.

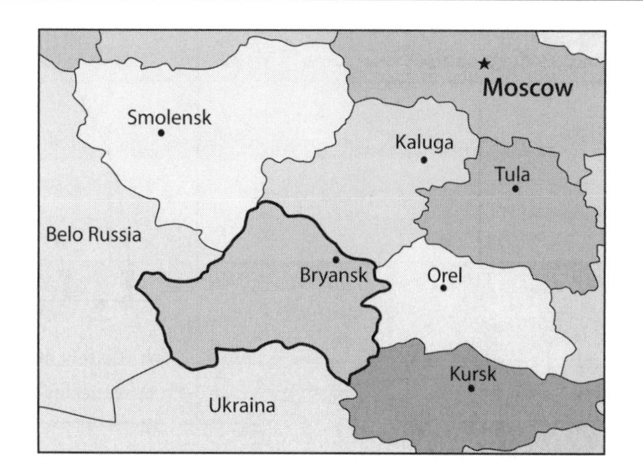

Fig. 5.1.
The Bryansk region in south-east Russia

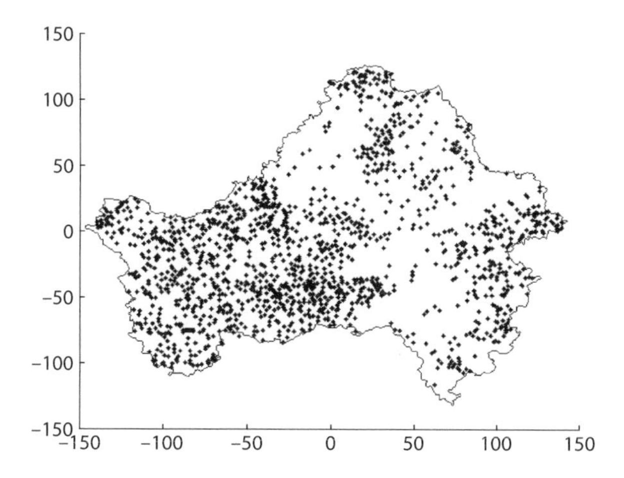

Fig. 5.2.
Locations of the monitoring stations over the Bryansk region (Russia)

where $C(\boldsymbol{u}, 1999)$ includes ^{137}Cs concentration measurements made at locations $\boldsymbol{u} \in V(\boldsymbol{s})$ during 1999, as well as projections of ^{137}Cs measurements made during previous years $t < 1999$ in the same subregion. One way to calculate these projections is in terms of the exponential decay model

$$C_V(\boldsymbol{s},1999) = C(\boldsymbol{s},t)e^{-(\ell n2 / T_{1/2})\tau} \tag{5.14}$$

where $\tau = 1999 - t$ and $T_{1/2}$ is the corresponding ^{137}Cs half-life. If, e.g., $C(\boldsymbol{u}, 1989)$ is a measurement made the year t=1989 at point $\boldsymbol{u}$ within $V(\boldsymbol{s})$, its projected value in year t=1999 is

$$C(\boldsymbol{u},1999) = C(\boldsymbol{u},1989)e^{-(\ell n2 / T_{1/2})\tau} \tag{5.15}$$

where $\tau = 10$ years (see also Fig. 5.3). To illustrate, soft data plots (in terms of probability densities f_S) obtained from a set of C_V concentrations (in Ci km^{-2}) at two of the sampling points (SP424 and SP482) in the Bryansk region (Russia) are shown in

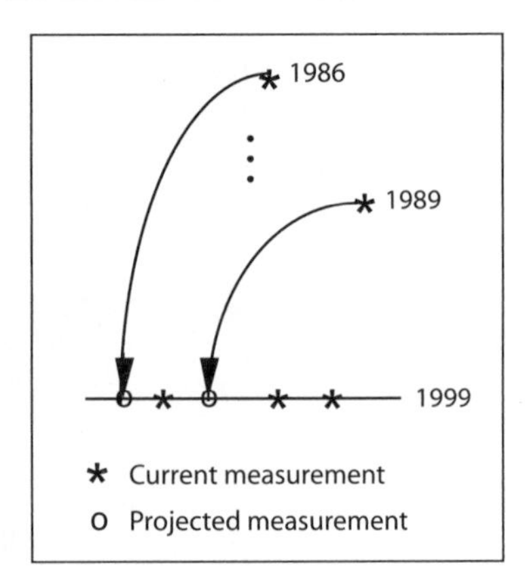

Fig. 5.3.
Current and projected ^{137}Cs measurements for the year 1999

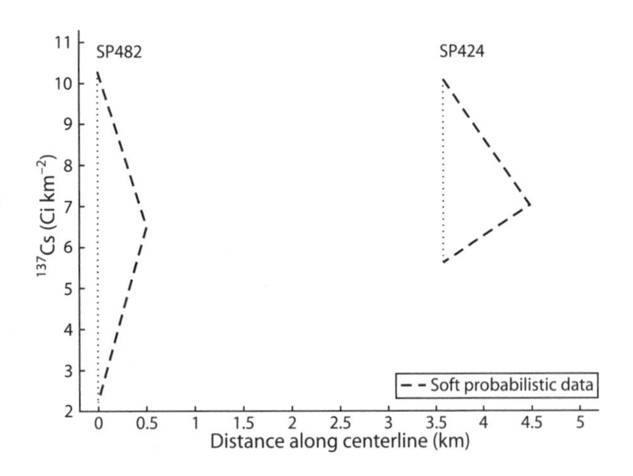

Fig. 5.4.
Examples of soft ^{137}Cs data ($Ci\ km^{-2}$) in the form of probability densities f_S obtained at two locations in the Bryansk region (Russia). The distance (in km) is measured along the centerline between the two points

Fig. 5.4. The different widths of the probability densities f_S at the two points reflect the different levels of uncertainty associated with the measurements obtained at these points. In the study of spatiotemporal variation, it is not isolated measurements but their relation to each other that is of interest. This relation is expressed by the variogram of the ^{137}Cs residual values (i.e., after the spatial trend has been subtracted from the C_V values at each point) along different directions. The spatial trend is due to the fact that, as a result of the various influencing factors at work (weather conditions, source terms, landscape, land use, etc.) the Chernobyl fallout affected unevenly different areas of Bryansk. In Fig. 5.5 a two-dimensional representation of the raw and the modelled variograms of the residuals are given. The shapes of these two-dimensional variograms reflect the distributional features of the ^{137}Cs concentrations and the subsurface characteristics of the region. On the basis of a series of similar types of soft data, the *bc*-based pdf and the *mb*-based pseudo-pdf were calculated for the year 1999 at a set of different locations in the Bryansk region. For illustration, the pdf and pseudo-pdf for the locations P1 through P4 are plotted in Fig. 5.6. The locations P1 and P4 coincide with the sampling points SP482 and SP424 of the monitoring network in

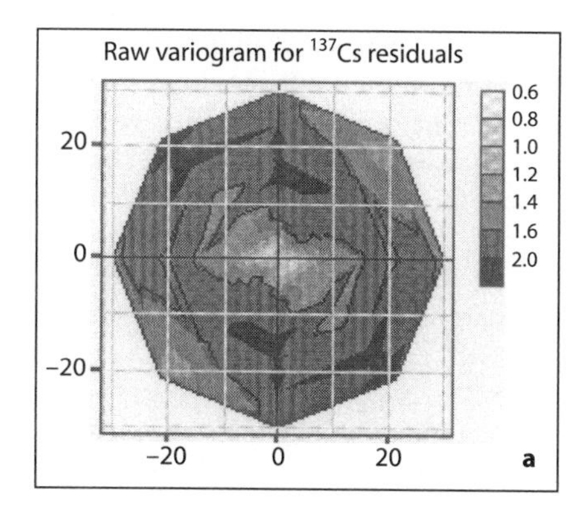

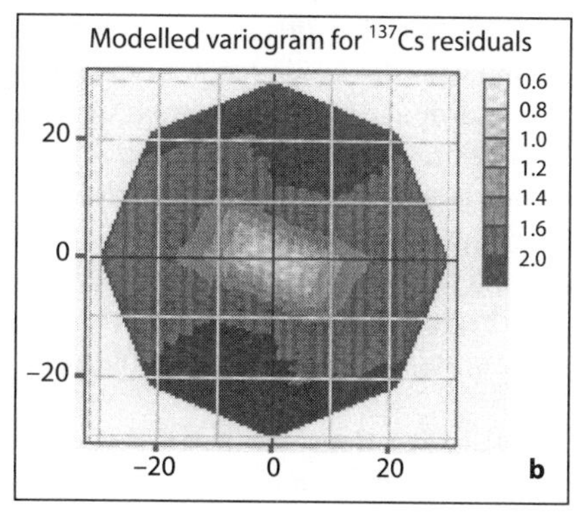

Fig. 5.5. **a** Raw and **b** modelled variograms of average ^{137}Cs values (in (Ci km^{-2})^2) for the year 1999

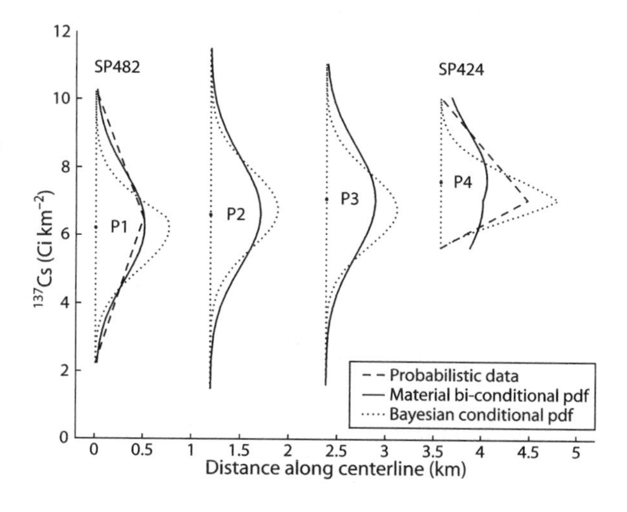

Fig. 5.6.
bc and *mb* pdf of ^{137}Cs values at 4 points in the Bryansk region during 1999. Soft (probabilistic) data at the 2 monitoring stations are also plotted

which soft data were available, whereas the locations P2 and P3 are situated outside along the centerline connecting the sampling points SP482 and SP424 (as shown in Fig. 5.6). As the reader might have expected, there are clear differences between the *bc*-based pdf and the *mb*-based pseudo-pdf at the above points. These differences can have some very important consequences. Indeed, they can lead to varying uncertainty assessments (e.g., $P_{\mathcal{K}}^{mb} \neq P_{\mathcal{K}}^{bc}$), different concentration estimates across space and time, etc. As a matter of fact, considerable differences were observed between the numerical ^{137}Cs concentration estimates at the points P1 through P4 derived on the basis of the *bc*-based pdf vs. the *mb*-based pseudo-pdf of Fig. 5.6. For instance, the *bc*-based mode ^{137}Cs concentration estimate at the point P3 was found to be equal to 6.78 Ci km^{-2}, whereas the *mb*-based mode estimate at the same point P3 was found to be equal to 7.09 Ci km^{-2}. Therefore, as these numerical investigations demonstrated, the *mb* approach can lead to different probabilistic structures and space/time maps than the *bc* approach. It is then up to the *TGIS* specialist to decide which approach offers a better representation of the physical situation under consideration. ▪

One interesting difference between the two forms of conditionalization is that, under certain conditions *mb* places more emphasis (than *bc*) on the contribution of the general knowledge in the derivation of the final (posterior) probability. Hence, *mb* could become more relevant in scientific situations in which well-established general knowledge (e.g., physical laws and theories) is available at the prior stage. On the other hand, the *bc*-based derivation of the posterior probability traditionally assigns extra weight to the site-specific knowledge (and, thus, it may be useful in applications where the *TGIS* expert has good reasons to put more emphasis on the site-specific data and less on an empirical law of the general *KB*).

Example 5.9. Consider the advection-reaction equation of pollutant distribution discussed in Example 6.8. Using *BME*, the physical law 6.26 and other forms of site-specific data were taken into consideration to generate the posterior pdf $f_{\mathcal{K}}^{bc}$ at various times and locations along the river. In the present example, the posterior *mb* pseudo-pdf $f_{\mathcal{K}}^{mb}$ is also calculated at several space/time points (mapping nodes). For illustration, Fig. 5.7 shows the *BME* pdf, the *mb* pseudo-pdf, as well as the pdf $f_{\mathcal{G}}$ of the prior stage calculated on the basis of the physical law 6.26. As was expected, the physical law-based $f_{\mathcal{G}}$ has a greater influence on the shape of $f_{\mathcal{K}}^{mb}$ than on that of $f_{\mathcal{K}}^{bc}$. The "peaked" shape of the latter is the result of the effect of the soft data (available at one neighboring point and the mapping node itself). As already mentioned, in such cases the expert may decide either in favor of *mb*, if he gives considerable priority to the physical law, or in favor of *bc*, if his judgment is that the emphasis should be given to the data. ▪

5.3.2 Material Conditionalization

Another non-Bayesian conditionalization approach may be constructed on the basis of *material conditionalization* (*mc*; denoted by →). The *mc* is based on a truth functional of the form $X_{\text{data}} \rightarrow X_k$, meaning that "it is not the case that the *S/TRF* event

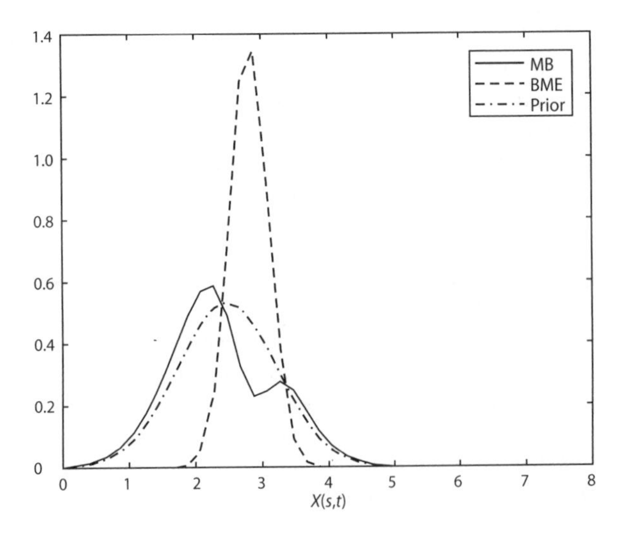

Fig. 5.7. *BME* pdf, *mb* pseudo-pdf, and prior pdf at a mapping grid node

X_{data} occurs and the X_k does not". Then, the *mc* probability law of the map may be defined by

$$P_{\mathcal{K}}^{mc}[x_k \leq \chi_k] = P_{\mathcal{G}}[X_{\text{data}} \rightarrow X_k] \quad (5.16)$$

where

$$P_{\mathcal{G}}[X_{\text{data}} \rightarrow X_k] = 1 - P_{\mathcal{G}}[X_{\text{data}}] + P_{\mathcal{G}}[X_{\text{data}}, X_k] \quad (5.17)$$

Interestingly, in this case it can be shown that one can mathematically define a *mc*-based pdf which has the same form as the *bc*-based pdf, although the corresponding probability laws are generally not the same, i.e., $P_{\mathcal{K}}^{mc} \neq P_{\mathcal{K}}^{bc}$ (which is clearly due to the different definitions of conditionalization that were used).

In a similar fashion to Sect. 5.3.1, if *mc* is used at the posterior (integration) stage of the space/time mapping procedure, the resulting modern geostatistical techniques include:

- The *McME* technique, i.e., *Material conditionalization Maximum Entropy.*
- The *McMF* technique, i.e., *Material conditionalization Maximum Fisherian* (epistemic sense) technique.

It is worth noticing that, the rich mathematical structure of the advanced *TGIS* functions makes possible the implementation of other conditionalization methods as well.

Finally, an interesting topic is the determination of mathematical conditions under which the non-Bayesian and the Bayesian conditionalization rules above give the same results. It turns out that for this to happen certain rather restrictive conditions need to apply. More specifically, the *bc* probability is equal to the *mb* probability if $P_{\mathcal{G}}[X_{\text{map}}]\{2 - P_{\mathcal{G}}^{-1}[X_{\text{data}}]\} = P_{\mathcal{G}}[X_{\text{data}}] + P_{\mathcal{G}}[X_k] - 1$. Furthermore, it can be shown that the *bc* probability is generally greater or equal to the *mc* probability, i.e. $P_{\mathcal{G}}[X_{\text{data}} \rightarrow X_k] \geq P_{\mathcal{G}}[X_k|X_{\text{data}}]$. The equality holds if and only if $P_{\mathcal{G}}[X_{\text{data}}] = 1$. In the case that $0 < P_{\mathcal{G}}[X_{\text{data}}] < 1$, the *bc* is equal to the *mc* if and only if $P_{\mathcal{G}}[X_k|X_{\text{data}}] = 1$. The epistemic interpretation of the above mathematical conditions may depend on the physical application under consideration.

5.4 By Way of a Summary

The *Modern Spatiotemporal Geostatistics* paradigm provides a *TGIS* framework that is conceptually well grounded, mathematically rigorous and philosophically sound. Nothing, however, is sacred in scientific research and development. Any framework should be open to possible revisions, when new knowledge makes it necessary (see also Sect. 8.5). Figure 5.8 presents some of the possible variants of the epistemic space/time mapping approach generated in the context of the *Modern Spatiotemporal Geostatistics* paradigm. These variants are combinations of the Bayesian and non-Bayesian (material) conditionalization rules with entropic and non-entropic (Fisherian) information measures. Hence, the *Modern Spatiotemporal Geostatistics* paradigm allows *TGIS* modelling to consider more than one information concepts and

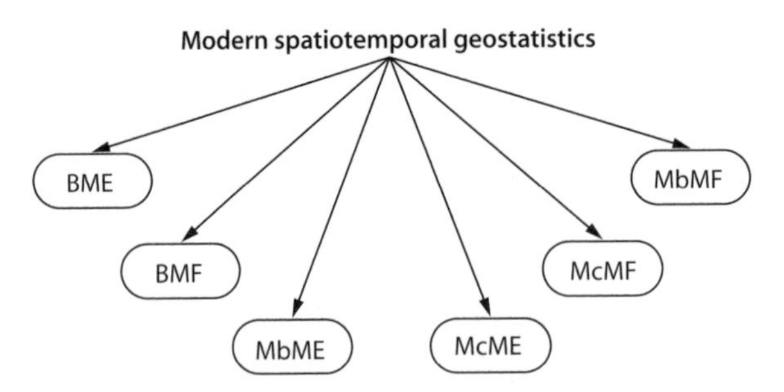

Fig. 5.8. Variants of the space/time mapping approach of *Modern Spatiotemporal Geostatistics*. *BME* = Bayesian Maximum Entropy, *BMF* = Bayesian Maximum Fisherian, *MbME* = Material biconditional Maximum Entropy, *McME* = Material conditional Maximum Entropy, *McMF* = Material conditional Maximum Fisherian, and *MbMF* = Material biconditional Maximum Fisherian

conditionalization principles, whenever the emerging conditions make it appropriate. However, we noted that compared to some other options currently under consideration, the information concept used in *BME* has certain important virtues. Furthermore, we saw that there is not a unique conditionalization rule that is universally valid in all cases, since no *a priori* (i.e., application-independent) reason exists which always favors one conditionalization rule over all others. Instead, the choice of a conditionalization approach should depend on the physical and logical characteristics of the specific application considered. Indeed, depending on the situation under study, the *TGIS* specialist may find it appropriate to construct alternative conditionalizations rules and then chose the one that fits the situation best. It was shown that the conditionalization rule used in *BME* has some advantages compared to classical Bayesianism (e.g., physical connections are taken into consideration). Surely, among all the methods shown in Fig. 5.8,

The BME method of space/time mapping is by far the best developed, at the moment. It has several appealing features, its analytical and computational properties are well understood, and it has been applied in a variety of TGIS applications with great success.

In Chap. 6, we will discuss some interesting numerical comparisons of Bayesian vs. non-Bayesian conditionalizations, which demonstrate the points made above. At the moment, it should be noted that Fig. 5.8 offers a convincing demonstration of the great opportunities for future research and development in the field of *TGIS*.

It was a time when even second-rate men did first-rate work,

once Paul Dirac remarked about a fruitful era of Physics in the mid 1920s. With this statement, his purpose was to emphasize the importance of working in a scientific field during a period when excellent conditions of growth, development and discovery exist. For many experts, this is the case of *TGIS* in nowadays.

Chapter 6

The *BME* Toolbox in Action

> But in science the credit goes to the man who convinces the world, not to the man to whom the idea first occurs.
> *F. Darwin*

6.1 The Fundamental *KB* Operators

The *TGIS* specialists know that a good set of tools, the knowledge to use them, and a sound theoretical background to support them are the three fundamental factors in getting a job done adequately and efficiently. Accordingly, a powerful *TGIS toolbox* which is the outcome of the theoretical *BME* approach will be discussed in this chapter.

As we saw in the preceding chapters, a novelty of the *BME* approach is that the indicators we rely on for measuring the performance of our mapping techniques should change from focusing solely on the results (i.e., statistically accurate estimates) to shedding light on the epistemic framework (logical rules, physical knowledge processing, etc.) we used to obtain those results so that we can influence map predictability in space/time. In brief, the spirit of the *BME* approach is to

Keep the derived maps consistent with the site-specific data while exactly satisfying the constraints arising from the physical laws.

In this way, the *BME* approach establishes a *unified* theoretical formalization that makes it possible to build a computationally effective framework for *TGIS* analysis and space/time mapping.

Like all scientific modellers, *TGIS* specialists need handy mathematical tools in order to work effectively. These tools are chosen so that they facilitate the job to be done and, in fact, they are indispensable if the job is to be done at all. In the following, the vectors $\boldsymbol{x}_{\text{map}}, \boldsymbol{p}_{\text{map}}, \boldsymbol{\chi}_{\text{map}}, \boldsymbol{\chi}_k$ and $\boldsymbol{p}_k$ refer to multi-point analysis[1]. From a computational viewpoint, it turns out that it is convenient to introduce two *KB*-related *operators*, as follows:

- The *general KB operator*

  $$y_{\mathcal{G}} = \boldsymbol{\mu}^T \boldsymbol{g} \tag{6.1}$$

  where the coefficients $\boldsymbol{\mu} = \{\mu_\alpha;\ \alpha = 1, \ldots, N\}$ depend on $\boldsymbol{p}_{\text{map}}$ (data and estimation points), and $\boldsymbol{g} = \{g_\alpha;\ \alpha = 1, \ldots, N\}$ are suitable functions of $\boldsymbol{\chi}_{\text{map}}$ which depend on the form of the $\mathcal{G}$-*KB* considered.

[1] The reader may recall that single-point analysis is always obtained as a special case of the multi-point formulation.

- The *specificatory KB operator*

$$\mathcal{Y}_S = \int_D d\Xi_S e^{\mu_0 + \mathcal{Y}_\mathcal{G}} \tag{6.2}$$

where $\mathcal{Y}_S$ is a function of the estimation points $\boldsymbol{p}_k$ and depends on the form of the $\mathcal{S}$-*KB* considered; the integrand $d\Xi_S$ and the domain D depend on the corresponding hard and soft data (see also Sect. 4.2.1).

At this point it is worth noticing that, *BME* analysis should not be carried out in isolation but in conjunction with other related *TGIS* activities (e.g., landscape characterization, physical modelling, or topographic surveying). A direct implication of this comment is the following thesis:[2]

A realistic implementation of BME in TGIS practice may require that different geographotemporal subdomains of the domain of interest be described in terms of different $\mathcal{Y}_\mathcal{G}$*- and* $\mathcal{Y}_S$*-operators.*

Therefore, the derivation of tractable analytical expressions for the operators $\mathcal{Y}_\mathcal{G}$ and $\mathcal{Y}_S$ is a very important part of the *BME* approach. Some examples will be given in Sect. 6.3 that follows.

6.2 Step-by-Step *BME*

6.2.1 The Formal Representation

In light of the preceding analysis, the basic *BME* equations of multi-point spatio-temporal mapping can be expressed in terms of the two *KB* operators, $\mathcal{Y}_\mathcal{G}$- and $\mathcal{Y}_S$, as follows,

$$\overline{h_\alpha}(\boldsymbol{p}_{\text{map}}) = \int d\boldsymbol{\chi}_{\text{map}} g_\alpha(\boldsymbol{\chi}_{\text{map}}) e^{\mu_0 + \mathcal{Y}_\mathcal{G}} \tag{6.3}$$

where $\alpha = 0, 1, \ldots, N$; and

$$f_\mathcal{K}^{bc}(\boldsymbol{\chi}_k) = A^{-1} \mathcal{Y}_S \tag{6.4}$$

Equation 6.4 offers a complete stochastic characterization of the map in terms of its multi-point pdf (A is the normalization coefficient that was defined in Sect. 4.2.1). The functions g_α ($\alpha > 0$) should be chosen so that the stochastic expectations $\overline{h_\alpha}$ in Eq. 6.3 can be calculated from the physical laws, empirical charts, etc. As usual, the $\overline{h_0} = g_0 = 1$ is a normalization constant. Equations 6.3 then express the *physical constraints* or *relationships* satisfied by the natural fields of the problem[3].

[2] This thesis is closely linked to the two golden rules of Sect. 3.4.2.

[3] Again, for more on the mathematical details involved in these formulas the reader is referred to Christakos ald Li (1998) and Christakos (2000).

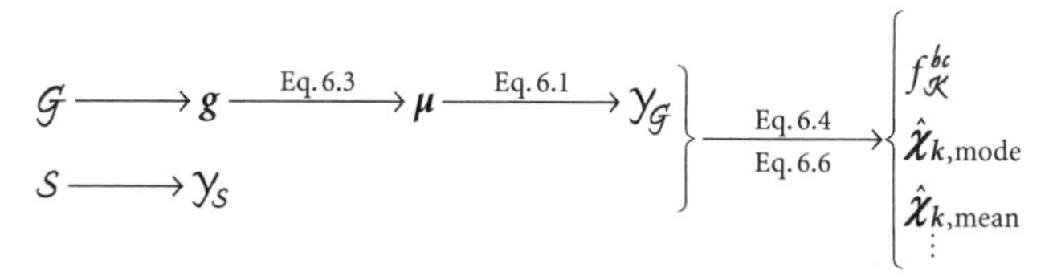

Fig. 6.1. Step-by-step *BME*

On the basis of the pdf $f_{\mathcal{K}}^{bc}$, the desired estimates $\hat{\chi}_k$ can be derived so that some optimality criterion is satisfied[4], thus leading to an estimated map for the entire spatiotemporal domain of interest,

$$\hat{\chi}_{\text{map}} = (\chi_{\text{data}}, \hat{\chi}_k) \tag{6.5}$$

Equation 6.5 is our best choice of a map χ_{map} *based* on the specified optimality criterion. Of course, $f_{\mathcal{K}}^{bc}$ offers a multiplicity of choices (mode, mean, median, etc. estimates) to the *TGIS* specialist. Often, the *TGIS* specialist seeks the most probable map, in which case one more basic *BMEmode* equation should be added to the basic *BME* equations above, namely,

$$\partial \mathcal{Y}_S / \partial \chi_{k_\ell} = 0 \tag{6.6}$$

at $\chi_k = \hat{\chi}_{k,\text{mode}}$, where $\ell = 1, \ldots, \rho$. Another popular choice of a map estimate is the *BMEmean* estimate, $\chi_k = \hat{\chi}_{k,\text{mean}}$, which minimizes the mean squared estimation error (for details, see Sect. 4.2.1).

Figure 6.1 summarizes schematically the five *basic steps* of the *BME* approach. These steps are as follows:

1. From the $\mathcal{G}$-*KB* available, formulate the vector $\mathbf{g}$. This vector automatically establishes the functional form of the $\mathcal{Y}_{\mathcal{G}}$-operator.
2. Substitute Eq. 6.1 into Eqs. 6.3, and solve for the vector $\boldsymbol{\mu}$.
3. Substitute $\boldsymbol{\mu}$ back into Eq. 6.1 to obtain the explicit form of the $\mathcal{Y}_{\mathcal{G}}$-operator.
4. From the $\mathcal{S}$-*KB* available, formulate the $\mathcal{Y}_S$-operator of Eq. 6.2.
5. Substitute operators $\mathcal{Y}_{\mathcal{G}}$ and $\mathcal{Y}_S$ into Eq. 6.4 and 6.6 to derive the $f\mathcal{K}bc$ and the $\hat{\chi}_{k,\text{mode}}$ map, respectively.

Following these steps, the multi-point pdf $f_{\mathcal{K}}^{bc}(\chi_k)$ of the life support field $X(\boldsymbol{p})$ under consideration can be derived, which exactly satisfies the physical constraints and honors the site-specific data. The derivation of the uni-modal $f_{\mathcal{K}}^{bc}(\chi_k)$ is a special case of the multi-point model.

Example 6.1. Fig. 6.2 illustrates a basic outcome of the *BME* approach above in the special case in which uni-modal pdf $f_{\mathcal{K}}^{bc}(\chi_k)$ have been derived at every point $\boldsymbol{p}_k = (s_k, t_k)$ of a space/time mapping grid in $R^1 \times T$. ▪

[4] For example, the estimates may be associated with a peak value of the pdf.

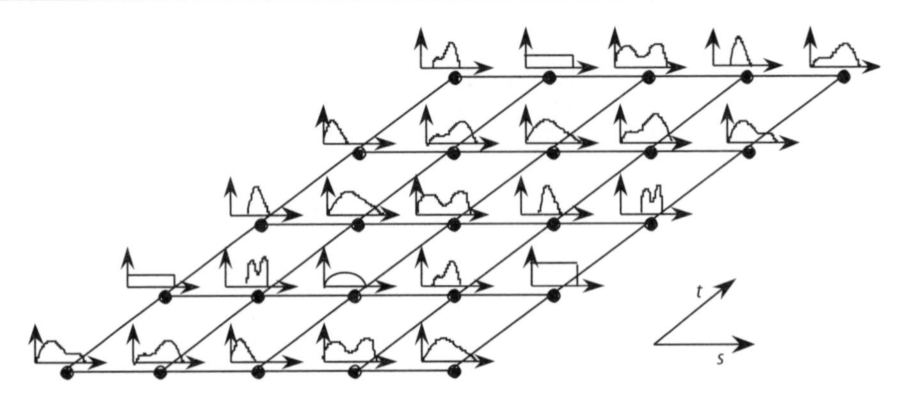

Fig. 6.2. *BME*-based pdf at every mapping point in the space/time domain $R^1 \times T$

6.2.2 The Diagrammatic Representation

Under certain circumstances, it may be useful to provide a *diagrammatic* representation of the *BME* approach. A detailed presentation of the diagrammatic approach can be found in Christakos and Hristopulos (1998). The basic idea of diagrammatic representation is that geometric objects (diagrams) can be used to represent mathematical expressions and to calculate them[5]. The diagrammatic representation provides a powerful visualization of the various space/time patterns, correlation components and multifold integrals incorporated by *BME*. Generally,

- A diagram consists of *objects* (e.g., lines and vertices), and represents mathematical formulas according to
- A set of *rules* that relate the objects to the corresponding mathematical formulas.

There is a one-to-one correspondence between diagrams and the various terms of a mathematical formula. For the diagrammatic representation of the various aspects of the *BME* approach the following set of rules may be useful:

Rule i	The $e^{y_\mathcal{G}}$ operator is represented in terms of small open circles (vertices) which denote the χ_i variables ($i = 1, 2, \ldots, m$) and are connected by thin straight lines.
Rule ii	The integration operation with respect to χ_i is represented by filling with a black color the corresponding open circle of the Rule (i) above.
Rule iii	Triangular arcs with p vertices ($p = 2, 3, \ldots$) denote multiplication of the corresponding χ_i variables.
Rule iv	Thick straight lines denote p-point statistics.

[5] Formal mathematics can be viewed as a language consisting of an alphabet of symbols and a set of operation rules that define how these symbols are combined. Diagrams can also be used for the same purpose, i.e., as tools for representing mathematical quantities.

The rules above are not the only ones possible. The reader can use another set of rules of his own choice. To illustrate the application of the specific rules above in a *BME* situation, let us consider a simple example.

Example 6.2. Assume that we have $m = 3$ data points ($\boldsymbol{p}_1$, $\boldsymbol{p}_2$, and $\boldsymbol{p}_3$), one estimation point $\boldsymbol{p}_k$, and the $\mathcal{G}$-*KB* consists of the (non-centered) covariances between these points. According to the rules (*i*) through (*iv*) above, a possible diagrammatic representation of the situation is as follows:

1 2 3 k $= e^{y_{\mathcal{G}}}$ (6.7)

The normalization constraint is diagrammatically represented as,

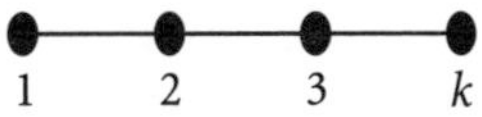

1 2 3 k $= \iiiint d\chi_1 d\chi_2 d\chi_3 d\chi_k e^{y_{\mathcal{G}}}$ (6.8)

and the two-point ($\boldsymbol{p}_1$ and $\boldsymbol{p}_2$) constraint is as follows,

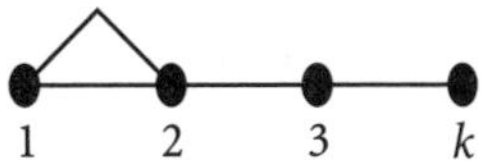

1 2 3 k $= \iiiint d\chi_1 d\chi_2 d\chi_3 d\chi_k \chi_1 \chi_2 e^{y_{\mathcal{G}}}$ (6.9)

If the (non-centered) covariance $C_x(\boldsymbol{p}_1,\boldsymbol{p}_2)$, e.g., is represented by

1 2 $= \overline{h_{12}}(\boldsymbol{p}_1,\boldsymbol{p}_2)$ (6.10)

the covariance-related constraints of Eqs. 6.3 are diagrammatically represented as follows:

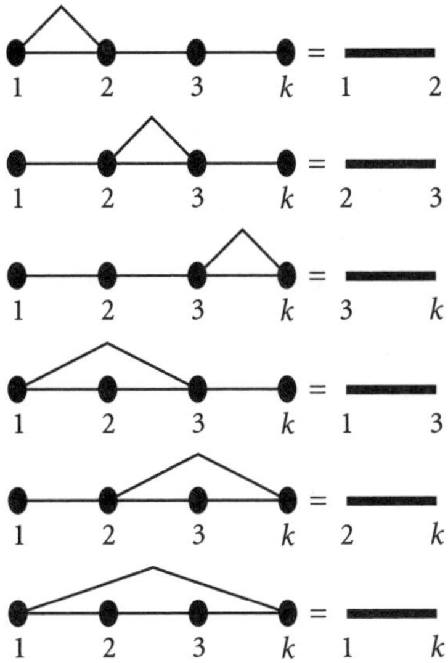

1 2 3 k = 1 2 (6.11a)

1 2 3 k = 2 3 (6.11b)

1 2 3 k = 3 k (6.11c)

1 2 3 k = 1 3 (6.11d)

1 2 3 k = 2 k (6.11e)

1 2 3 k = 1 k (6.11f)

Constraints 6.3 involving higher order statistics can be represented diagrammatically in a similar fashion. Furthermore, assume that the $\mathcal{S}$-*KB* available consists of two hard data at points $\boldsymbol{p}_1$ and $\boldsymbol{p}_2$ and an interval (soft) datum at point $\boldsymbol{p}_3$. Then, in light of the aforementioned rules, the posterior pdf $f_{\mathcal{K}}^{bc} = A^{-1}\int_I d\chi_3 e^{\mathcal{Y}_{\mathcal{G}}}$ is diagrammatically expressed as follows:

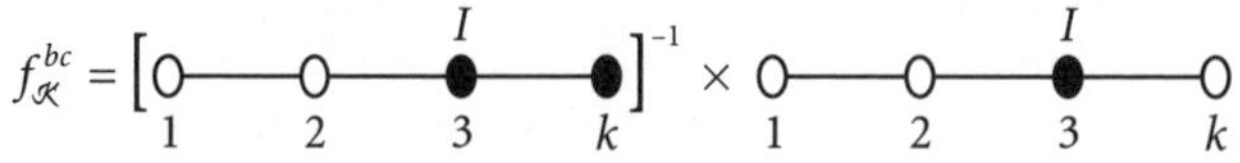

$$f_{\mathcal{K}}^{bc} = \Big[\underset{1}{\circ} \text{---} \underset{2}{\circ} \text{---} \overset{I}{\underset{3}{\bullet}} \text{---} \underset{k}{\bullet}\Big]^{-1} \times \underset{1}{\circ} \text{---} \underset{2}{\circ} \text{---} \overset{I}{\underset{3}{\bullet}} \text{---} \underset{k}{\circ} \tag{6.12}$$

where I indicates the restricted domain of integration of the interval soft values χ_3. ▪

Diagrammatic representation provides a compact notation for *BME* analysis, and can help evaluate the relative importance of the different terms in *BME* calculations. By means of a backward application of the diagrammatic rules, analytical equations are obtained. Hence,

Diagrammatic representation offers a simple but at the same time instructive model as to how BME works.

When they make themselves familiar with it, many readers could find the diagrammatic representations easier to use than the formal representation in terms of mathematical symbols[6].

In the previous chapters the reader was exposed to the mathematical and epistemological principles underlying the *BME* approach. In the following sections we focus on a number of examples, in an attempt to provide practical demonstrations that the *BME*-based technology can improve the quality of *TGIS* analysis and space/time mapping. The main goal of this chapter is didactic, i.e. to make the reader familiar with the *BME* toolbox, including

- The basic *BME* concepts.
- The step-by-step analytical and computational techniques.
- The use of the computer library *BMElib* (see also the following Chap. 7).

Therefore, some of the examples discussed in the following sections of this chapter are analytical, whereas several others are essentially numerical experiments based on synthetic (simulated) datasets (case-studies which refer to real-world situations have been presented in previous chapters as well as in the relevant literature). Each one of these two groups of examples has its own advantages. The analytic studies make it possible for the readers to be exposed to the basic conceptual steps and analytical formulas of the *BME*-based techniques. The numerical experiments, on the other hand, provide a controlled environment so that the readers can concentrate on certain crucial elements of the computational techniques (without having to take into account cumbersome details and interpretations that are unavoidable when dealing with real datasets), and allow them to obtain insightful comparisons of *BME* performance vs. other *TGIS* mapping techniques. Moreover, with the help of these numerical experiments

---

[6] The philosophically inclined reader may draw a parallel between the "diagrammatic representation" above and Wittgenstein's "picture theory" of language (Wittgenstein 1961).

readers can familiarize themselves with the *BMElib*[7]. Although *BME* can use a wide variety of general knowledge sources, for computational convenience as well as for comparing *BME* with other techniques, we assume in most of this chapter's examples that the general knowledge consists of the mean and covariance (or variogram) functions.

6.3 Analytical and Numerical Case-Studies

In this section, we will illustrate the various uses of the *BME* toolbox with the help of a few (analytic and synthetic) examples and case-studies. The reader can reproduce many of the numerical examples with the help of the *BMElib* software included for this purpose.

6.3.1 Some Commonly Encountered Situations

We start with a few commonly encountered situations in *TGIS* practice, in which the general *KB* involves 2nd-order space/time statistics, and the site-specific *KB* consists of hard and soft data.

Example 6.3. Let us follow the steps of the systematic *BME* procedure of Sect. 6.2 above. In steps 1, 2 and 3 we assume that the $\mathcal{G}$-*KB* consists of the mean vector $\overline{x_{\text{map}}}$ (multi-point case), and the (centered) ordinary covariance matrix c_{map} between all pairs of points. Then, the solution of Eqs. 6.3 leads to the $\mathcal{Y}_{\mathcal{G}}$-operator shown in Eq. 6.13 of Table 6.1.

Table 6.1. Examples of common $\mathcal{Y}_{\mathcal{G}}$-operators

$\mathcal{G}$	$\mathcal{Y}_{\mathcal{G}}$	
$\overline{x_{\text{map}}}$ c_{map}	$-0.5(\boldsymbol{\chi}_{\text{map}} - \overline{x_{\text{map}}})^T c_{\text{map}}^{1} (\boldsymbol{\chi}_{\text{map}} - \overline{x_{\text{map}}})$	(6.13)
γ_{map}	$-0.25\sum_{i=k1}^{k_\rho}\left[\sum_{j=1}^{m}\gamma_{ij}^{-1}(\chi_i - \chi_j)^2 + \sum_{i=k1}^{k_\rho}\gamma_{ij}^{-1}(\chi_i - \chi_j)^2\right]$	(6.14)
$dX(t)/dt = bX(t)$	$\chi_k \sigma_x^{-2}(0)\left[\overline{X(0)} - 0.5\chi_k e^{-bt_k}\right]e^{-bt_k}$	(6.15)

Table 6.2. Examples of $\mathcal{Y}_S$-operators

S	$\mathcal{Y}_S[\cdot]$	
Eq. 3.18	$\int_I d\boldsymbol{\chi}_{\text{soft}}[\cdot]$	(6.16)
Eq. 3.19	$\int_I dF_S(\boldsymbol{\chi}_{\text{soft}})[\cdot]$	(6.17)
Eq. 3.20	$\int_{R^1} dF_S(\zeta;h)\int_{I(\zeta)} d\boldsymbol{\chi}_{\text{soft}}[\cdot]$	(6.18)

[7] Our hope is that these examples are properly chosen so that they highlight the considerable potential of the *BME* approach, including its superiority over many of the existing mapping techniques (space/time regression, Kriging, objective analysis, etc.).

In step 4 we let the $\mathcal{S}$-*KB* include hard data of the form of Eq. 3.15 and soft (probabilistic) data of the form of Eq. 3.18. Then, the $\mathcal{Y}_S$-operator is as in Eq. 6.17 of Table 6.2. In step 5, and in view of the previous steps, the integration pdf is given by

$$f_{\mathcal{K}}^{bc}(\boldsymbol{\chi}_k) = A^{-1}\int_I dF_S(\boldsymbol{\chi}_{\text{soft}})e^{\mu_0+\mathcal{Y}_\mathcal{G}} \tag{6.19}$$

Furthermore, the *BMEmode* vector estimates $\hat{\boldsymbol{\chi}}_k$ are solutions of the multi-point equations ($\ell = 1, \ldots, \rho$)

$$\sum_{i=1}^{m_h} c_{ik_\ell}^{-1}(\chi_i - \overline{x_i}) + \sum_{i=m_h}^{m} c_{ik_\ell}^{-1}\left[\overline{x_i}(\hat{\boldsymbol{\chi}}_k) - \overline{x_i}\right] + \sum_{i=k_1}^{k_\rho} c_{ik_\ell}^{-1}(\hat{\chi}_i - \overline{x_i}) = 0 \tag{6.20}$$

where $c_{ik_\ell}^{-1}$ is the ik_ℓ-th element of the inverse matrix $\boldsymbol{c}_{\text{map}}^{-1}$, $\overline{x_i}(\hat{\boldsymbol{\chi}}_k) = \mathcal{Y}_S[\chi_i e^{\mathcal{Y}_\mathcal{G}}] \,/\, \mathcal{Y}_S[e^{\mathcal{Y}_\mathcal{G}}]$ at $\boldsymbol{\chi}_k = \hat{\boldsymbol{\chi}}_k$. ■

Example 6.4. Assuming that the general *KB* includes the variogram matrix $\boldsymbol{\gamma}_{\text{map}}$, and by following the same procedure as above, we get a $\mathcal{Y}_\mathcal{G}$-operator of the form of Eq. 6.14 in Table 6.1. If the site-specific *KB* includes hard data of the form of Eq. 3.15 and soft (interval) data of the form of Eq. 3.18, the $\mathcal{Y}_S$-operator is of the form 6.16 in Table 6.2. Then, the integration pdf is given by

$$f_{\mathcal{K}}^{bc}(\boldsymbol{\chi}_k) = A^{-1}\int_I d\boldsymbol{\chi}_{\text{soft}} e^{\mu_0+\mathcal{Y}_\mathcal{G}} \tag{6.21}$$

and the multi-point *BMEmode* estimate $\boldsymbol{\chi}_k$ is the solution of the set of equations ($\ell = 1, \ldots, \rho$)

$$\sum_{j=k_1}^{k_\rho} \gamma_{k_\ell j}^{-1}(\hat{\chi}_{k_\ell} - \chi_j) + \tfrac{1}{2}\sum_{j=1}^{m_h} \gamma_{k_\ell j}^{-1}(\hat{\chi}_{k_\ell} - \chi_j) + \tfrac{1}{2}\sum_{j=m_h+1}^{m} \gamma_{k_\ell j}^{-1}\overline{x_j}(\hat{\chi}_{k_\ell}) = 0 \tag{6.22}$$

where γ_{ij}^{-1} is the ij-th element of the inverse matrix $\boldsymbol{\gamma}_{\text{map}}^{-1}$, $\overline{x_j}(\hat{\chi}_{k_\ell}) = \mathcal{Y}_S[(\hat{\chi}_{k_\ell} - \chi_j)e^{\mathcal{Y}_\mathcal{G}}] \,/\, \mathcal{Y}_S[e^{\mathcal{Y}_\mathcal{G}}]$ at $\boldsymbol{\chi}_k = \hat{\boldsymbol{\chi}}_k$. ■

Note that in the last case of Table 6.1, the general *KB* includes a temporal physical model, where b is a deterministic parameter. From the physical model we derive the corresponding $\mathcal{Y}_\mathcal{G}$-operator as shown in Eq. 6.15, where $\overline{X(0)}$ and $\sigma_x^2(0)$ are the mean and variance, respectively, at the time origin. Due to space limitations, theoretical formulations for only a few selected (general and specificatory) *KB* have been considered here. However, there are almost infinite possibilities, limited only by the availability of such physical *KB* in the *TGIS* applications. For instance, several other examples may include combinations of the $\mathcal{Y}_\mathcal{G}$- and $\mathcal{Y}_S$-operators of Tables 6.1 and 6.2.

6.3.2 Spatiotemporal Filtering

This section is concerned with space/time *filtering*, i.e., the case where soft data are available at the estimation points themselves. A computationally efficient formulation of the *BME* equations in this case is summarized in the following example (see also a relevant discussion in the numerical Example 4.5).

Example 6.5. Consider the case of single point mapping, in which the mapping points $p_{map} = (p_{hard}, p_{soft}, p_k)$ include the hard data points p_{hard}, the soft data points p_{soft}, and the estimation point p_k. The $\mathcal{G}$-*KB* is expressed in terms of the mean vector $m_{map} = \overline{x_{map}}$, and the covariance matrix $c_{map} = \overline{(x_{map} - m_{map})(x_{map} - m_{map})^T}$. The $\mathcal{S}$-*KB* includes hard data χ_{hard} at p_{hard}, and soft data of the probabilistic type at p_{soft}, as well as soft data of the same type at the estimation points, see Eq. 3.21, so that $f_S^{(ks)}(\chi_{soft}, \chi_k) = f_S(\chi_{soft}) f_S^{(k)}(\chi_k)$. Without loss of generality, we assume that $m_{map} = 0$, and let $c_{a,b} = \overline{(x_a - m_a)(x_b - m_b)^T}$ be the covariance matrix between sets of points a and b; e.g., the $c_{hard,k} = \overline{(x_{hard} - m_{hard})(x_k - m_k)^T}$ denotes the covariance matrix between p_{hard} and p_k. Similarly, we define the conditional covariance vector and matrix as $B_{a|b} = c_{a,b} c_{b,b}^{-1}$ and $c_{a|b} = c_{a,a} - B_{a|b} c_{b,a}$, respectively. In order to simplify notation, we also denote the n-variate Gaussian pdf (m, c) as follows,

$$\phi(\chi; m, c) = (2\pi)^{-n/2} |c|^{-1/2} \exp\left[-0.5(\chi - m)^T c^{-1} (\chi - m)\right]$$

Then, the computationally efficient *BME* equations used to calculate the posterior pdf, the *BMEmean* estimate, and the associated error variance are, respectively (Serre et al. 1998; Serre 2001),

$$f_{\mathcal{K}}^{bc}(\chi_k) = A^{-1} f_S^{(k)}(\chi_k) \phi(\chi_{kh}; 0, c_{kh,kh}) \int d\chi_{soft} f_S(\chi_{soft}) \phi(\chi_{soft}; B_{soft|kh} \chi_{kh}, c_{soft|kh}) \quad (6.23)$$

$$\overline{x}_{k|\mathcal{K}} = (A')^{-1} \int d\chi_{ks} \chi_k f_S^{(ks)}(\chi_{ks}) \phi(\chi_{ks}; B_{ks|hard} \chi_{hard}, c_{ks|hard}) \quad (6.24)$$

$$\sigma_{k|\mathcal{K}}^2 = (A')^{-1} \int d\chi_{ks} (\chi_k - \overline{x}_{k|\mathcal{K}})^2 f_S^{(ks)}(\chi_{ks}) \phi(\chi_{ks}, B_{ks|hard} \chi_{hard}, c_{ks|hard}) \quad (6.25)$$

where $\chi_{kh} = (\chi_{hard}, \chi_k)$, $\chi_{hs} = (\chi_{hard}, \chi_{soft})$, $A = \phi(\chi_{hard}; 0, c_{hard,hard}) A'$, and $A' = \int d\chi_{ks} f_S^{(ks)}(\chi_{ks}) \phi(\chi_{ks}, B_{ks|h} \chi_{hard}, c_{ks|hard})$. ▪

6.3.3 Exogenous Information

The data sources considered in Sect. 6.3.1 and 6.3.2 may be related to expert knowledge about a specific site, previous expertise with similar situations, or some other kind of empirical relation between a (well-observed) *exogenous* natural field and the geographical field of interest. A *TGIS* expert may have to his possession, e.g., preexisting maps of an exogenous (continuous or categorical) field over the area of interest. As it turns out, in most cases

The exogenous field provides vague although valuable information that can be formally incorporated into TGIS by means of the BME formalism.

Depending on the scientific discipline and application considered, the exogenous information can take several forms – the simplest one being interval data. More elaborate forms involve probability distributions at the estimation points. Below we present two numerical examples that explore these possibilities. The numerical experiments performed in these examples assume a set of hard data for the geographical field of interest (say, X_1), as well as a good amount of information in terms of an exogenous field (say, X_2); X_2 is spatially correlated with X_1 (e.g., if X_2 represents altitude data from a digital elevation model and X_1 denotes temperature sampled at a set of points, the two fields may be related by means of an altimetric negative gradient-based model). Then, on the basis of knowledge about X_2, one can define a useful conditional pdf for the field X_1 at the estimation nodes.

Example 6.6.[8] Assume that the $X_1(s)$ and $X_2(s)$, $s = (s_1, s_2)$, are (0, 1)-Gaussian random fields which vary over a square area D of unit size. Their spatial variations are homogeneous and are characterized by exponential variograms (their ranges and sills are all equal to 1; in appropriate units). Their cross-variogram is, also, exponential (its range is equal to 1 and its sill is equal to 0.9). The point correlation between X_1 and X_2 is $\rho = 0.9$ (Fig. 6.3a). Additional site-specific knowledge includes the conditional pdf, $f_x(\chi_1|\chi_2 > 0)$ and $f_x(\chi_1|\chi_2 < 0)$; see Fig. 6.3b (the pdf are assumed available at any point over D). Note that these pdf are not Gaussian (they are negatively and positively skewed). Realizations of X_1 and X_2 were generated over D, using the Cholesky decomposition simulation technique (for a detailed discussion of this technique, see Christakos 1992). In Fig. 6.4a a spatial map showing a X_1 realization is plotted (this simulated map will serve as the actual map, for the purposes of the present analysis). Simulated values of the X_1 realization at 25 locations over D were considered as hard data (Fig. 6.4b), and a simple Kriging technique was first applied to estimate X_1 at 400 points on a 20×20 grid covering the area D. The derived estimation map is denoted as *KH* (denoting Kriging with hard data) and is plotted in Fig. 6.4.c. Next, the simulated X_2-values at the 400 estimation points were replaced by the appropriate conditional X_1 pdf, depending on whether the X_2-values are above or below zero. The map of the mean values (*MV*) of these conditional pdf at the 400 points is then plotted in Fig. 6.4d. These pdf served as soft data at the estimation points for the *BME* technique (i.e., in addition to the 25 hard data above), thus leading to the map of the *BMEmean* estimates plotted in Fig. 6.4e. A careful examination of the mapping results displayed in Fig. 6.4c

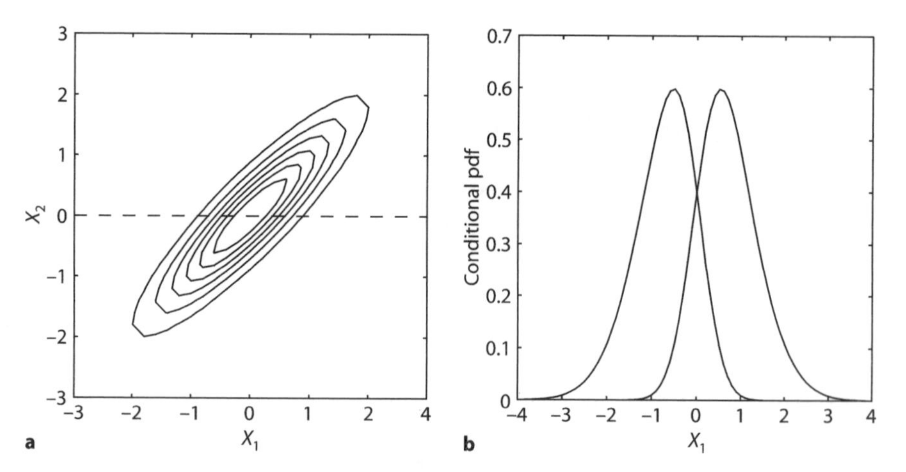

Fig. 6.3. a Isodensity curves for bivariate Gaussian pdf ($\rho = 0.9$); **b** conditional pdf for X_1 given that X_2 is below zero (*left curve*) or above zero (*right curve*)

[8] The reader can use the routine ***example01.m*** to reproduce the results of this example.

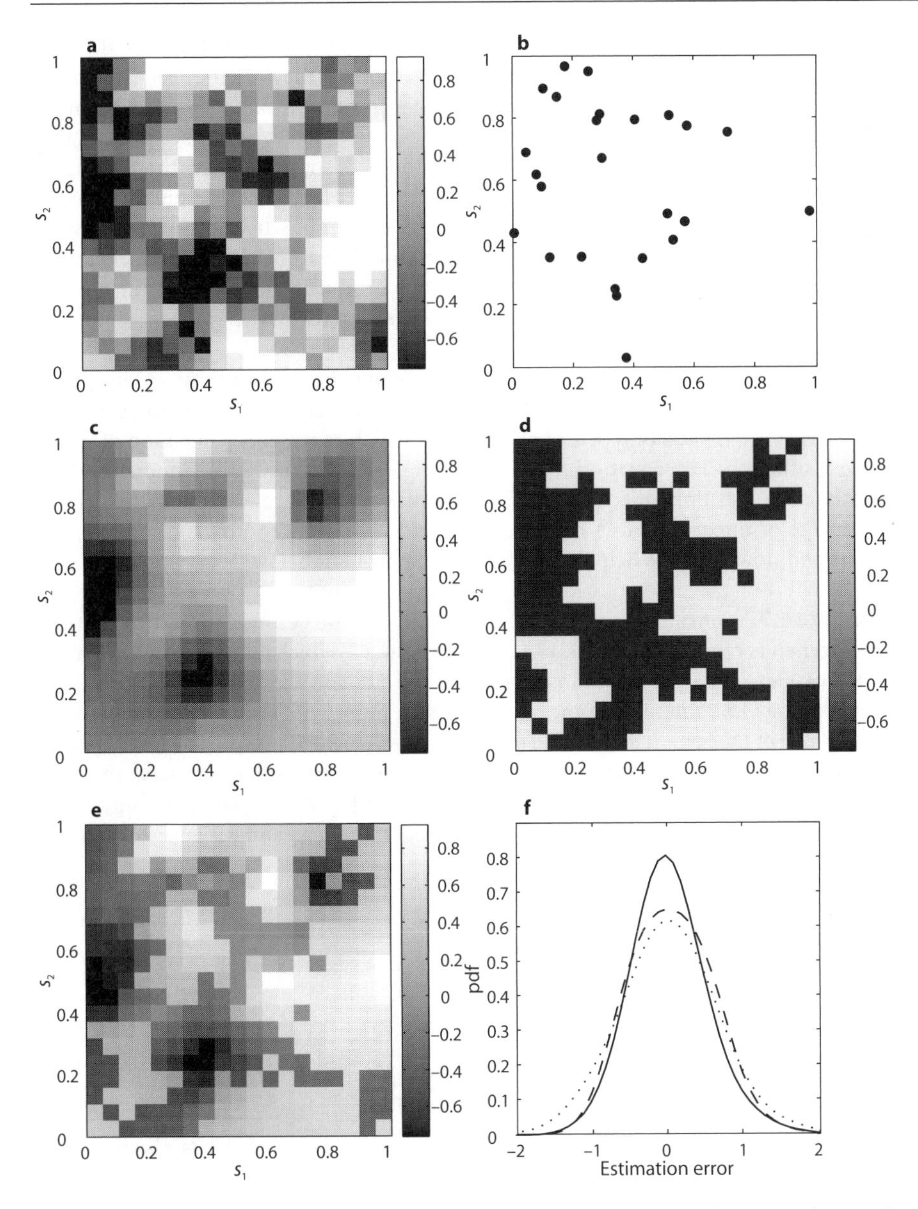

Fig. 6.4. a Simulated X_1 map over the square area D; **b** locations of the 25 hard data; **c** *KH*, **d** *MV*, and **e** *BMEmean* maps of X_1; **f** error distributions for *KH* (*dotted line*), *MV* (*dashed line*) and *BME* (*plain line*)

through 6.4e reveals the mapping improvements gained as a result of taking into account the exogenous soft information. Indeed, the conclusion reached was that,

By rigorously taking into account the exogenous information, BME *does a much better job than either* KH *or* MV *in representing the spatial pattern of the actual map.*

(High-valued and low-valued spatial variations, e.g., are better represented by the *BME* map than by the *KH* or the *MV* maps.) As a matter of fact, the *BME* map seems to properly combine the most important features of the *KH* and the *MV* maps. The *KH* map, on the other hand, is far too smooth and does not adequately reproduce the variability of the simulated values at the estimation nodes. Also, statistically the *BME* technique performs better than the *KH* and *MV* techniques (Fig. 6.4f), despite the considerably uncertain information provided by the exogenous field. Even if the mean squared error for *BME* (i.e., 0.19) is not much lower than that for *MV* (i.e., 0.22), by allowing smoother variations the *BME* technique produces a map that is much closer to reality (as represented by the map of Fig. 6.4a) than the *MV* map. ▪

There exist real-world applications in which, while there is not a one-to-one correspondence between the observed values of the natural or the epidemiological fields, these fields are related in terms of an equation which represents some kind of an averaged situation. The equation may express, e.g., an empirical or phenomenological relationship that links the field of interest with one or more exogenous fields. The advantage of integrating this kind of site-specific knowledge into *TGIS* modelling is illustrated next with the help of a numerical experiment below.

Example 6.7.[9] Consider the square area D of the previous example, and two spatially homogeneous random fields $X_1(\boldsymbol{s})$ and $X_2(\boldsymbol{s})$, $\boldsymbol{s} = (s_1, s_2)$, which take their values over D. In the case examined in this example, though, the site-specific knowledge includes the empirical fact that the value of the field X_2 at each point $\boldsymbol{s}$ of D is, on the average, equal to the value of the field X_1 at the same point plus a zero mean Gaussian and spatially independent error term with standard deviation equal to 1/4 times the X_1 value at the same point (the error term is considered heteroskedastic; Fig. 6.5). Then, at each point $\boldsymbol{s}$, a Gaussian conditional pdf $f_x(\chi_1|\chi_2 = \chi_{2,k})$ is defined with mean and variance equal to $\chi_{2,k}$

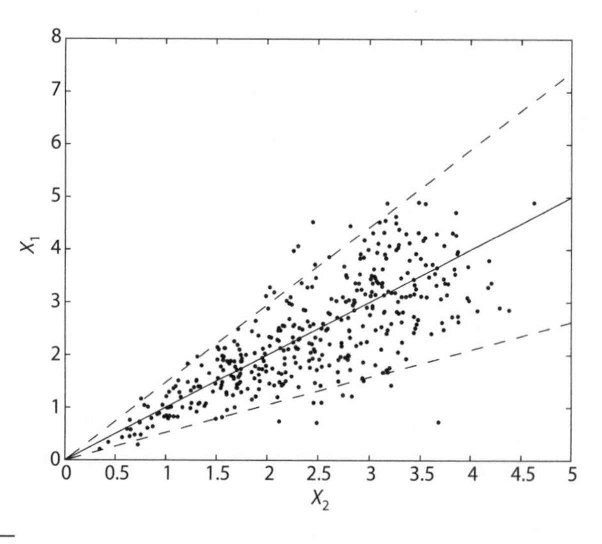

Fig. 6.5. Expected linear relationship between X_1 and X_2 (*plain line*) together with the 0.025 and 0.975 quantiles of the conditional Gaussian pdf of X_1 given X_2 (*dashed lines*). The 400 simulated values of X_1 and X_2 are superimposed on the graph

[9] The reader can use the routine ***example02.m*** to reproduce the results of this example.

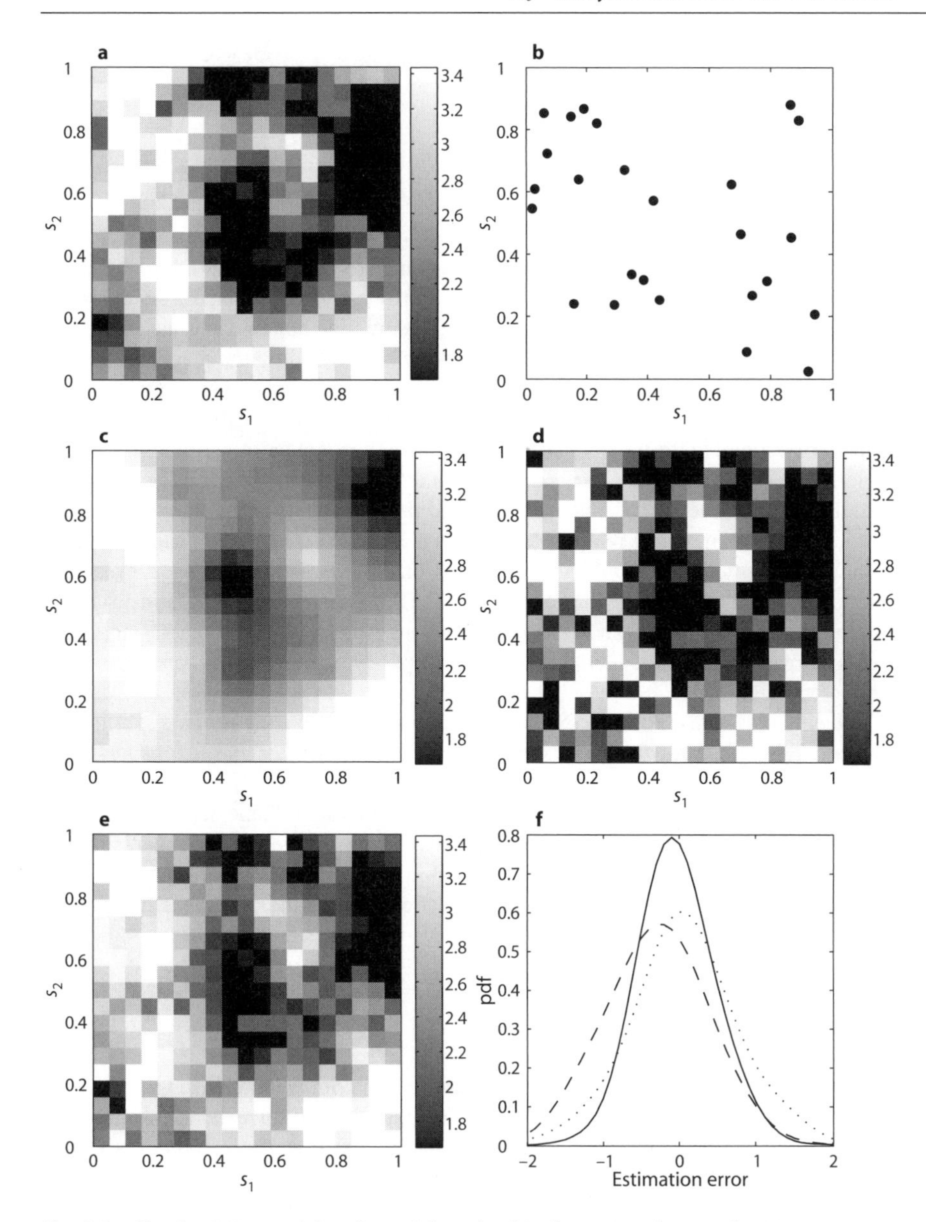

Fig. 6.6. a Simulated X_1 map; **b** locations of the 25 hard X_1 data; **c** *KH*, **d** *MV*, and **e** *BMEmean* maps of X_1; **f** error pdf for *KH* (*dashed line*), *MV* (*dotted line*), and *BME* (*plain line*)

and $(1/16)\chi_{2,k}$, respectively. Using the same random field simulation technique as in the previous Example 6.6, a spatial realization of a (2.5, 1) Gaussian field X_1 with exponential variogram (its range and sill are both assumed equal to 1) was generated over the area D (Fig. 6.6a). For the purpose of this example, the X_1 values at the 25 points shown in Fig. 6.6b were considered as hard data. To each one of the 400 simulated values of the X_1

Table 6.3.
Estimation errors

	Mean error	Mean squared error
KH	-0.30	0.48
MV	-0.06	0.45
BME	-0.04	0.17

field at the estimation nodes, a zero mean Gaussian error term with standard deviation as above was added in order to obtain the map of the $\chi_{2,k}$ values. Using these values, the conditional pdf of X_1 given X_2 were found at each estimation node s_k. As usual, in the case of ordinary Kriging only the 25 hard X_1 data were considered by the mapping scheme. Following the notation of the previous example, the corresponding estimation map is denoted as *KH* and is plotted in Fig. 6.6c. Furthermore, the map of the *MV*-based estimates (which correspond to the mean values of the conditional pdf at the 400 estimation points) is plotted in Fig. 6.6d, for comparison. On the other hand, the *BME* technique rigorously accounts for both the hard X_1 data and the conditional pdf data at the estimation nodes provided by the empirical relationship between X_1 and X_2. The *BMEmean* map thus obtained is plotted in Fig. 6.6e. An examination of the results (Fig. 6.6c–e) should convince the reader about the improvement in the *BME* estimation of the field X_1 gained as a result of the soft data derived from the X_2 values. Not surprisingly, the *KH* technique yields an overly smoothed spatial map with only a few details. On the contrary,

By being able to use site-specific conditional pdf, the BME map exhibits spatial patterns of high- and low-valued areas that are more realistic than those of the KH map.

Furthermore, as the statistics displayed in Fig. 6.6f and in Table 6.3 amply demonstrate, the *BME* estimation errors (i.e., the mean errors as well as the mean squared errors) are smaller than both the corresponding *MV* estimation errors and the *KH* estimation errors (as a matter of fact, the reader may notice that these differences turned out to be quite substantial). ▪

6.3.4 Physical Laws

The following example is a synthetic case-study in a controlled environment, which allows the reader to compare the *BME*-based results with the corresponding analytical solutions of a physical law. Furthermore it demonstrates how *BME* solution is able to incorporate other kinds of soft information which the usual solution process cannot.

Example 6.8. Consider the stochastic advection-reaction equation of pollutant distribution along a one-dimensional space and in time, as follows (e.g., Spitz and Moreno 1996),

$$-v\frac{\partial}{\partial s}X(s,t) - kX(s,t) = \frac{\partial}{\partial t}X(s,t) \tag{6.26}$$

where $X(s,t)$ is the pollutant concentration (in ppm), v denotes the velocity (in km day^{-1}), and k is the reaction rate constant (in day^{-1}). Suppose that Eq. 6.26 represents the time distribution of pollutant concentration along a river, and that at some location $s = 0$

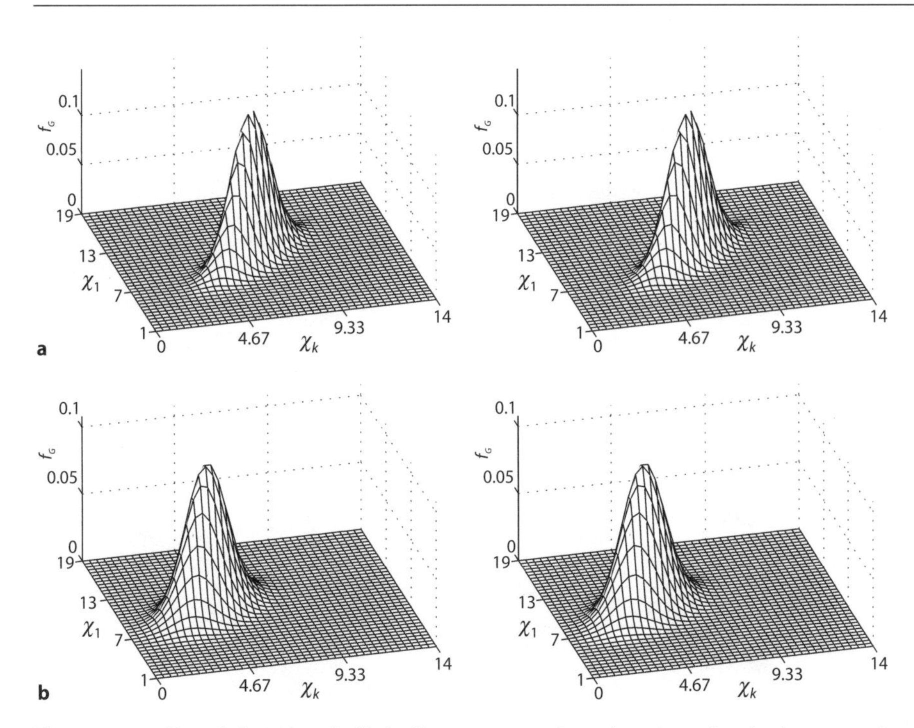

Fig. 6.7. General knowledge $\mathcal{G}$-based pdf of pollutant concentration at the points: **a** $(s_k, t_k) = (0.409, 0.001)$, and **b** $(s_k, t_k) = (0.82, 0.001)$. The analytical solutions are shown for comparison; $k = 0.25$ day^{-1}

and time $t = 0$ the pollutant concentration is X_0. An analytical solution of the law 6.26 is

$$X(s,t) = X_0 \exp[(cv - k)t - cs] \tag{6.27}$$

where X_0 is the initial pollutant concentration, and $c = \varepsilon^{-1}$ (ε is a characteristic spatial length; in km). For the purpose of the present analysis, it is assumed that X_0 is modelled by a Gaussian random variable with mean $\overline{X_0} = 10$ ppm and variance $\sigma_0^2 = 2$ ppm^2. The velocity of the river flow is $v = 10$ km day^{-1} and the reaction rate constant is $k = 0.25$ day^{-1}, which means that the pollutant concentration decreases in time. The coefficient c is, also, considered as a random variable with mean $\overline{c} = 1.05$ and variance $\sigma_c^2 = 0.08$. Using the *BME* approach, the advection-reaction law 6.26 is integrated into the space/time analysis leading to the $\mathcal{G}$-based bivariate pdf $f_{\mathcal{G}}(\chi_1, \chi_k)$ between an initial point, say $(s_1, t_1) = (0.001, 0.0001)$, and several other space/time points (s_k, t_k) along the river (Kolovos et al. 2000). In Fig. 6.7a and b we display these pdf for various space/time distances $r_{k1} = |s_k - s_1|$ and $\tau_{k1} = t_k - t_1$. In all the cases examined above,

An excellent agreement between the numerical BME solutions of the physical law and the analytical solutions was found.

(the analytical solutions are also plotted in Fig. 6.7a and b, for comparison). Furthermore, in Fig. 6.8a and b the pdf plots are shown in the case where the reaction rate

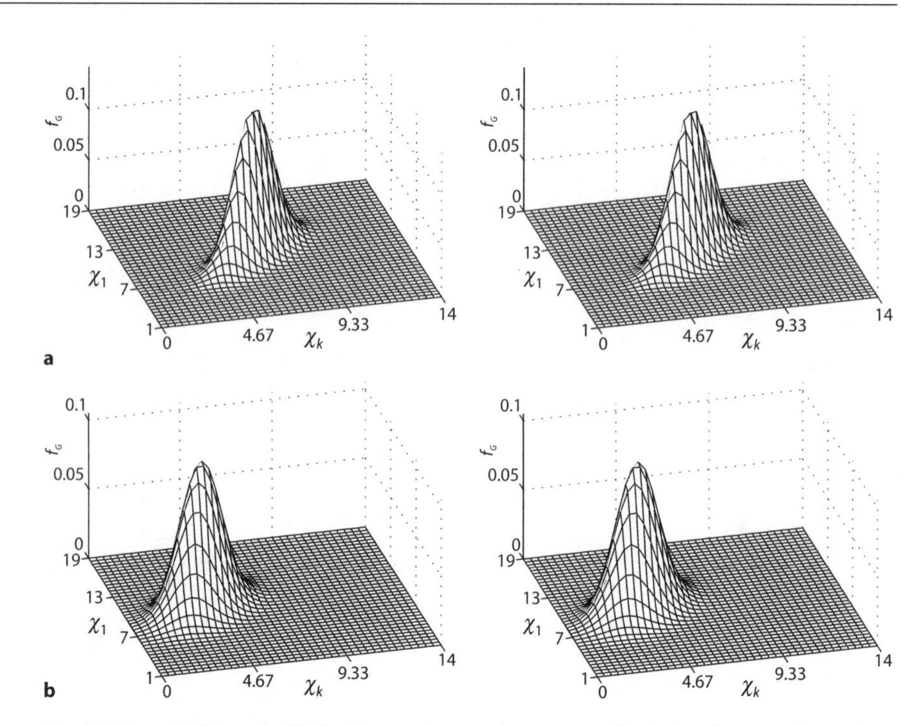

Fig. 6.8. General *KB*-based pdf of pollutant concentration computed by *BME* at the space/time points, **a** $(s_k, t_k) = (0.425, 0.001)$, and **b** $(s_k, t_k) = (0.95, 0.001)$. Analytical solutions are shown for comparison; $k = 0.5$ day^{-1}

constant was increased by 100% (i.e., $k = 0.5$ day^{-1}). Again, in all these cases the numerical *BME* results are in excellent agreement with the analytical solutions. Next, at the meta-prior *BME* stage, soft data of the interval and the probabilistic types were generated on the basis of a total of 5 000 pollutant concentration values, which were simulated using Eq. 6.27.

For illustration, we assumed first that soft data of the interval type and then soft data of the probabilistic type were available at the space/time point $(s_1, t_1) = (0.001, 0.0001)$. Then, the integration (or posterior) pdf, $f_{\mathcal{K}}^{bc}(\chi_k)$, of the pollutant concentrations were computed at five spatially equidistant estimation points (s_k, t_k) along the river, i.e. $ds = 0.125$ km, and corresponding to the same time $t_k = 0.0001$ days. Similarly, pollutant concentration pdf can be calculated at any other space/time point and for an arbitrary number of data points. These five pdf, together with the corresponding interval and probabilistic data that were used by the *BME* approach, are displayed in Fig. 6.9a and b, respectively. The reader may notice the varying shapes of the pollutant concentration pdf at the different estimation points along the river. In light of these insights, we concluded that another important advantage of the *BME*-based analysis of the physical law of Eq. 6.26 is that,

The final BME solutions account not only for the stochastic physical law and its initial/boundary conditions but for many other kinds of site-specific information as well.

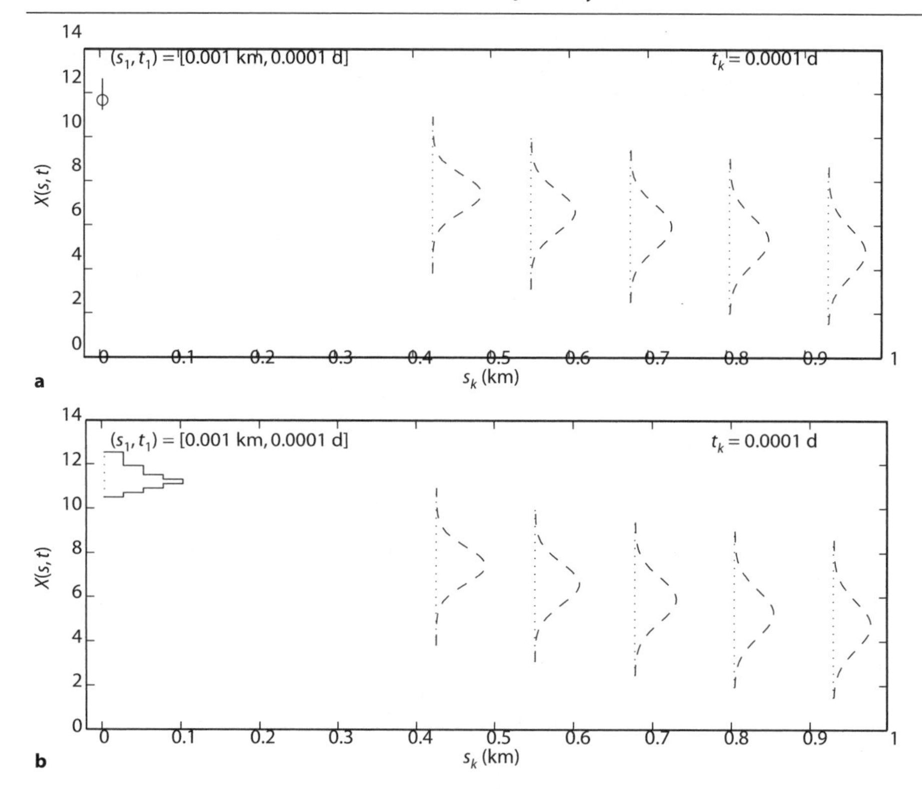

Fig. 6.9. Integration (or posterior) pdf of pollutant concentration at a set of estimation locations along the river and at a specified time instant. The bar and pulse in **a** and **b** indicate, respectively, the interval and probabilistic datum point used in *BME* analysis

(Including the soft data of both the interval and the probabilistic types mentioned above.) The *BME*-based study of the physical law 6.26 has several other appealing features. One of these features, e.g., is the fact that,

Instead of a single realization/solution of the pollutant distribution, the complete posterior pdf of the pollutant concentrations are derived at all space/time points of interest.

To the authors' knowledge, no other composite space/time mapping technique (including statistical regression, neural networks, perturbation, expansions, and spectral analysis) possesses these powerful features of the *BME* approach. ▪

6.3.5
Using Soft Data to Improve *TGIS* Mapping

In this section, the advantage of using soft data in space/time mapping is further illustrated with the help of two numerical experiments. These experiments have been designed so that they study the relative effects of fixed vs. moving data points. In particular, hard data are assumed available at a number of fixed locations surrounding the estimation point, whereas a single interval datum is located at varying distances from the estimation point. It is assumed that the general knowledge consists of the mean and covariance (or variogram) functions. Then, as should be intuitively expected, the difference between the Kriging estimation error (obtained using only hard data) and the *BME* estimation error (obtained using hard and interval data) increases as the moving interval datum approaches the estimation point. Moreover, the benefit of using soft information increases as the width of the interval datum gets smaller, reflecting improved information quality.

Example 6.9.[10] Assume that 5 hard data points of a spatially homogeneous geographical random field $X(s)$, $s = (s_1, s_2)$, are arbitrarily located over a square area D of unit size. The estimation point is located at the center of D. As is shown in Fig. 6.10a, there is, also, one interval soft datum at a point moving along the diagonal connecting the upper left corner of D with its center. In order to quantify the relative advantage of using the moving interval datum in the mapping process, 400 sets of simulated values are jointly generated at the estimation point, at the 5 hard data points, and at the 3 possible locations of the moving interval datum. The simulated field $X(s)$ (considered as the actual field, for the purposes of this example) is (0,1)-Gaussian with an exponential variogram (range $\varepsilon = 1$ and sill $c_0 = 1$; in appropriate units). The moving interval datum has a constant width $\Delta\chi = 0.5$, with randomly chosen lower and upper limits so that the simulated value is on the average uniformly distributed within the interval. Using, in turn, each one of the 3 possible interval datum locations of the moving point, the *BMEmode* estimation error distributions are derived (Fig. 6.10b and Table 6.4). In particular, *BME*/1, *BME*/2, and *BME*/3 correspond to the largest, intermediate, and smallest (moving data point-estimation point) distance, respectively. The Kriging estimates obtained using only the 5 hard data (as usual, denoted by *KH*) are also calculated, for comparison. Table 6.4 shows that for the most remote location of the

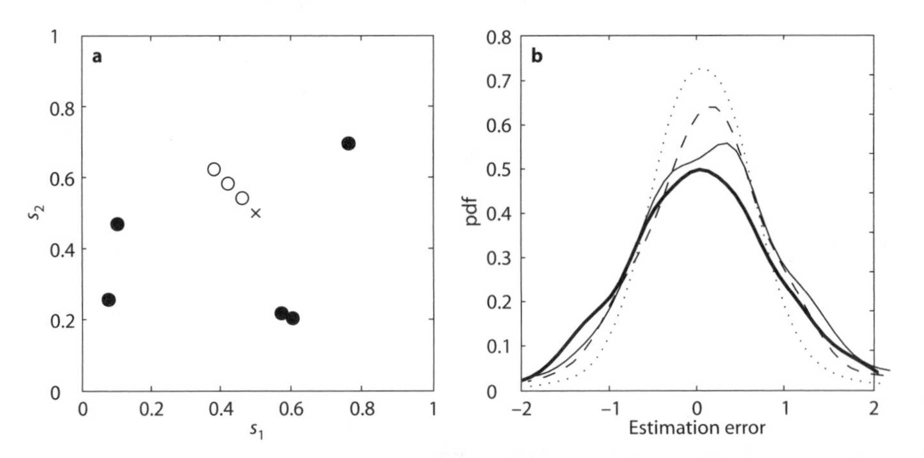

Fig. 6.10. a The 5 fixed locations of the hard data (*black spots*), the 3 possible locations of the interval datum (*white circles*) and the estimation point (*cross*). **b** Estimation error distributions of Kriging (*bold line*) and *BMEmode* for the 3 possible interval datum locations (BME/1 in *plain line*; BME/2 in *dashed line*; BME/3 in *dotted line*)

[10] The reader can use the routine ***example03.m*** to reproduce the results of this example.

Table 6.4.
Kriging and *BME* estimation errors

	Mean error	Mean squared error
KH	0.03	0.67
BME/1	0.02	0.52
BME/2	0.02	0.43
BME/3	−0.01	0.25

interval datum, there is little difference between *KH* and *BME/1* estimation errors. A more noticeable difference occurs at the intermediate (*BME/2*). Finally, at the smallest distance the mean squared error of *BME/3* is less than one-half that of *KH*. Not surprisingly, the conclusion derived from the above remarks is that,

Integrating soft data into the mapping process can be very valuable, especially when the data points are close enough to the estimation point.

Since soft data have a lower information content than hard data, including soft data into the mapping scheme will have a marked effect mainly in areas where relatively few hard data are available. ▪

Another significant issue is concerned with the effect of the varying *widths* of the interval data on the mapping results. This effect is highlighted in the following example.

Example 6.10.[11] We assume the same arrangement as in Example 6.9 above, with the difference that now the location of the interval datum is fixed (Fig. 6.11a) and its width varies from 0.1 to 4.3 units. Using the same technique as before, 1 000 spatial realizations

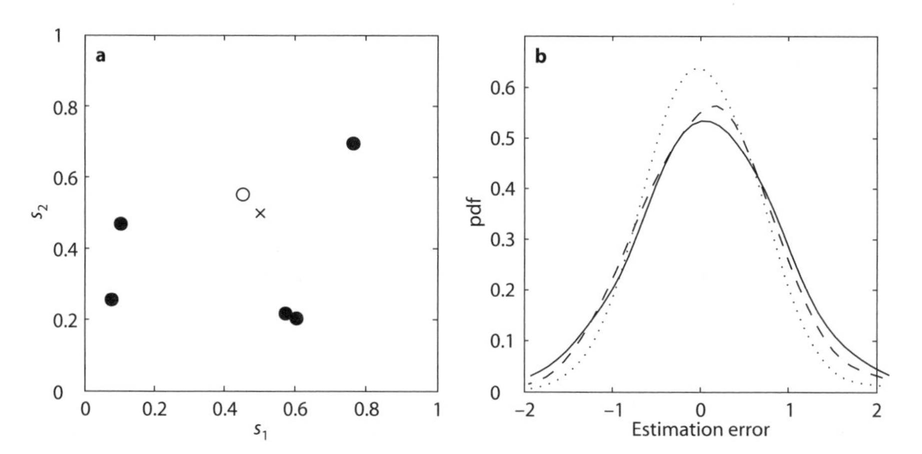

Fig. 6.11. a Locations of the 5 hard data (*black spots*), the 1 interval datum (*white circle*) and the estimation point (*cross*). **b** *BMEmode* estimation error distribution for varying interval datum width (BME/1 in *plain line*; BME/2 in *dashed line*; BME/3 in *dotted line*)

[11]The reader can use the routine ***example04.m*** to reproduce the results of this example.

Table 6.5.
Kriging and *BME* estimation errors

	Mean error	Mean squared error
Ka	–0.02	0.68
BME/1	–0.02	0.55
BME/2	–0.03	0.46
BME/3	–0.03	0.32
Kb	–0.03	0.31

were simulated over the area D. The estimates obtained by three mapping methods were then compared. The first method is simple Kriging based on 5 hard data (denoted as *Ka*), the second one (denoted as *Kb*) is simple Kriging using 6 hard data (i.e., the same 5 values at the hard data points and the simulated value at the location of the interval datum), whereas the third method (*BME*) uses the 5 hard data as well as the additional interval datum with various widths and generates *BMEmode* estimates. These estimates (denoted as *BME*/1, *BME*/2, and *BME*/3) correspond to interval datum widths of 4.3, 2.2, and 0.1 units, respectively. The estimation error distributions of the three *BME* variants above are plotted in Fig. 6.11b. Moreover, as is shown in Table 6.5, the estimation errors obtained by the *BME* technique lie between the errors obtained using the *Ka* and the *Kb* techniques. (Note that in this case, *Kb* is equivalent to *BME* with 6 hard data only.) The larger the width of the interval datum is, the closer the *BME* results are to those of *Ka*. On the other hand, for small interval widths the *BME* technique yields results very similar to those of *Kb*. This situation emphasizes the consistency of the *BME* approach, i.e.,

BME properly accounts for the information quality of the soft data as expressed by the width of the relevant intervals.

When this quality level is low (e.g., the interval width is large), the soft datum plays little or no role in estimation, and *BME* produces similar results to Kriging with no soft data. On the other hand, when the soft information quality is high (e.g., the interval width is very small), soft data have almost the same effect on *BME* estimation as would have hard data at the same locations. ▪

Taking a clue from the insights of the preceding numerical experiments, in the next example we consider a situation (very common in soil mapping) in which one needs to draw maps of a spatially or spatiotemporally *continuous* property, using only soft data-based knowledge (i.e., very few or no accurate measurements are usually available in these cases; see also Sect. 4.2.3).

Example 6.11.[12] Consider 400 locations over a 20×20 regular grid covering a square area D of unit size. At these locations, the *categories* to which the values belong are known and are the only (soft) information available. The aim is to produce a continuous-valued estimation map over D on the basis of the soft information. To avoid methodological issues, we assumed that the theoretical variogram model is

[12] The reader can use the routine ***example05.m*** to reproduce the results of this example.

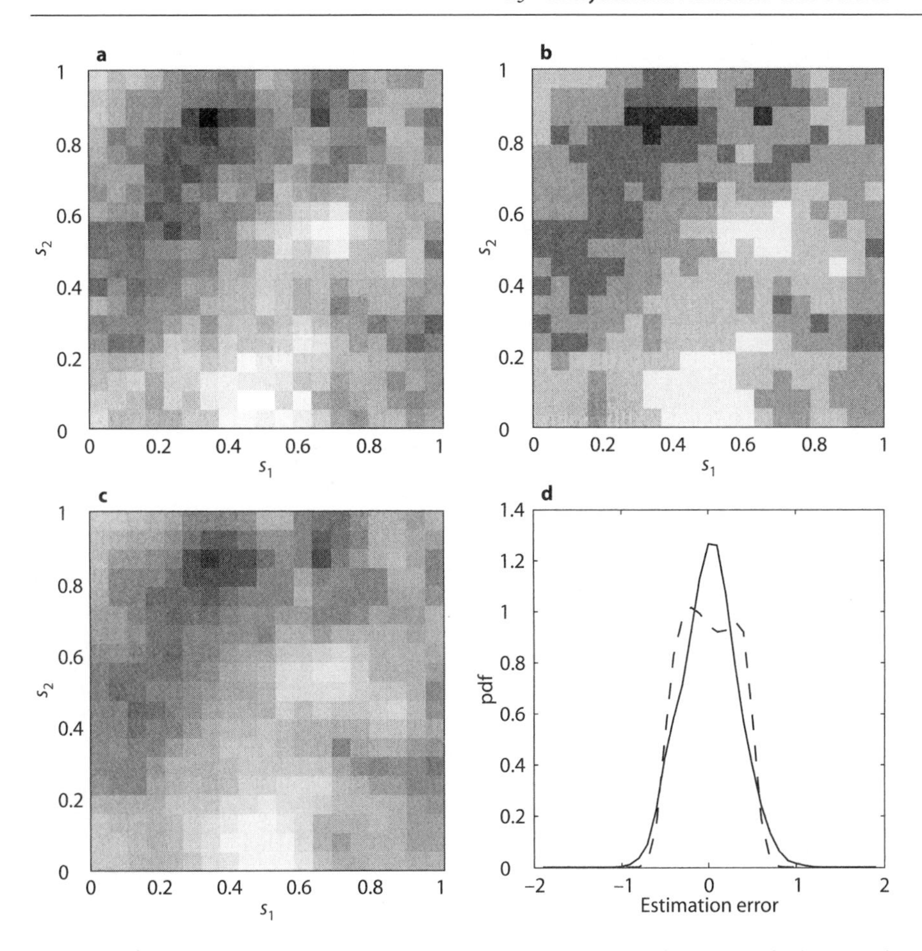

Fig. 6.12. a Simulated map; **b** and **c** estimation maps for *MI* and *BMEmode*, respectively (lowest and highest values are coded as black and white, respectively); **d** kernel smoothed distribution of the prediction errors for *BME* (*plain line*) and *MI* (*dashed line*)

known (in practice, the variogram must be estimated from the categories, e.g., by using indicator coding of the intervals). In Fig. 6.12a values of a $(0,1)$ Gaussian random field $X(s)$, $s = (s_1, s_2)$, are generated at 400 estimation nodes using a simulation technique as in previous examples with a spherical variogram model (its range is equal to 0.5 and its sill is equal to 1). The 400 simulated hard data at the sampling locations are then replaced by the corresponding unit width interval data. The bounds of these interval data are as follows: $-3, -2, \ldots, 2, 3$. In order to assess the performance of the *BME* technique under these circumstances, we will compare it to another technique that simply considers the middle of the interval datum as the estimated value (*MI*) at each point. In the view of many experts, this kind of a simplistic approach is, in fact, the only alternative to the *BME* technique, in this case. A technique like Indicator Kriging (*IK*; see Sect. 4.3.2; also, see later, Sect. 6.5.2) which uses a binary coding for each class is useless in this case, since *IK* is an exact interpolator and it will simply

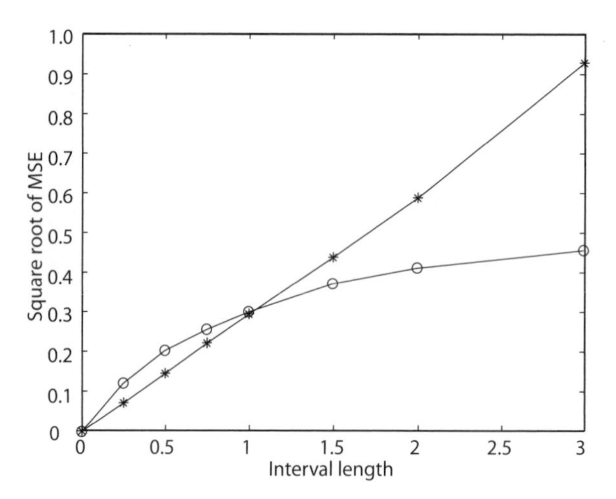

Fig. 6.13.
Evolution of the mean squared error (*MSE*) root as a function of the interval width for *MI* (*stars*) and *BME* (*circles*)

provide the same intervals as the soft (categorical) data at the mapping points. Figure 6.12b shows that the *MI* technique produces "patches" of identical values, since no variation of values is allowed inside the same interval. Since the exact values are almost uniformly distributed over the intervals, the *MI* estimation error distribution (Fig. 6.12d) is almost uniform over the interval [–0.5, 0.5]. On the other hand, the *BMEmode* map (Fig. 6.12c) exhibits smooth variations, in close agreement with the reference map (Fig. 6.12a). The *BME* approach yields a Gaussian estimation error distribution centered around zero (Fig. 6.12d). In conclusion, on the basis of a categorical map consisting of only 6 classes of values, *BME* was able to generate a good reproduction of the reference map. The key reason for *BME*'s success to reconstruct efficiently the continuous-valued reference map from the corresponding categorical data-based map is the fact that,

BME does not only process the local soft interval data at each estimation point, but it also accounts for the manner in which these intervals are spatially arranged.

This valuable spatial information is neglected by *MI*, which is a rather *ad hoc* technique lacking any sound theoretical basis. It is interesting to study how the estimation error distribution changes as a function of the interval width. For this purpose, various interval widths ranging from 0 to 3 units were considered. Both mapping techniques are unbiased but, while for the *MI* technique the increase in the mean squared error is linear, for the *BME* technique it shows an asymptotic behavior (see Fig. 6.13). Clearly, the maximum rate of increase in the *BME* estimation errors occurs for small interval data widths, whereas the rate of increase tends to slow down considerably for larger widths. This is intuitively justified (e.g., doubling the interval width from 0.25 to 0.50 should have a much greater effect on the mean squared error than increasing the width from 2.75 to 3.0). On the contrary, in the case of the *MI* technique the mean squared error increases linearly, which is rather counter-intuitive and highlights a weakness of the particular technique when processing soft datasets. ▪

6.3.6
Non-Bayesian Analysis

As we saw in Chap. 5 in certain physical situations, implementing a *non-Bayesian* approach (e.g., material biconditionalization, *mb*) instead of a *Bayesian* conditionalization (*bc*) method at the posterior (integration) mapping stage may lead to different *TGIS* results. We also suggested that these differences are worth studying. For numerical illustration, some plots of *bc*-based pdf and *mb*-based pseudo-pdf are presented in the example below.

Example 6.12. Various simulations were generated of a spatiotemporal random field $X(\boldsymbol{p}), \boldsymbol{p} = (\boldsymbol{s}, t)$, with a zero mean and a space/time covariance of the exponential form

$$c_x(r,\tau) = 80e^{-3r/3.6-3\tau/15} \tag{6.28}$$

where the $r = |s_2 - s_1|$ and $\tau = t_2 - t_1$ denote the spatial and temporal lags, respectively. We assumed that interval (soft) data were available at the points $\boldsymbol{p}_1$ and $\boldsymbol{p}_2$, which are located at a spatial distance Δr units apart and occur at the same time instant t. The estimation point $\boldsymbol{p}_k$ was located in the middle of the spatial distance Δr, also at time t. Different widths, $\Delta\chi$, were assumed for the interval data above, and different Δr values were used in the numerical investigations. For illustration, some of the posterior densities obtained are plotted in Fig. 6.14 On the basis of the numerical experiments the conclusion was that,

The mb-based pseudo-pdf of a space/time can be considerably different than the bc-based pdf of the map.

One obvious difference, e.g., was that the *mb*-based pseudo-densities tend to take a larger variety of shapes (both symmetric and asymmetric) than the *bc*-based densities. It was also found that, as the interval width $\Delta\chi$ increases, the *mb*-based pseudo-

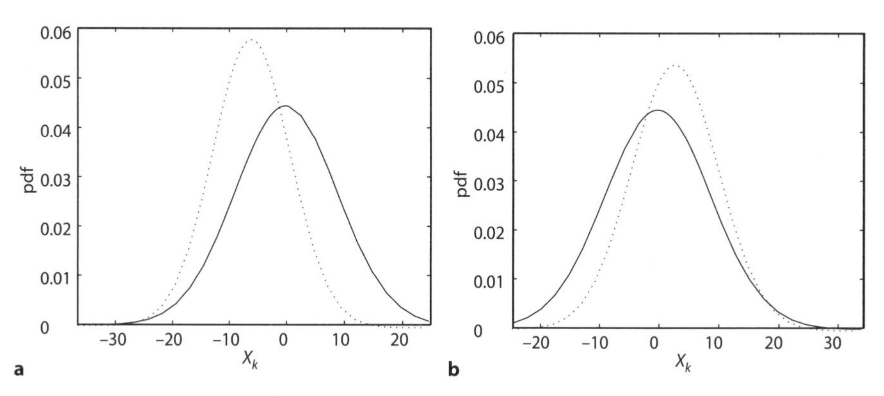

Fig. 6.14. Pdf obtained from *bc* (*dotted lines*) and *mb* (*solid lines*): **a** data intervals, [−9.10, −4.62] and [−9.85, −5.38]; $\Delta r = 1.58$; **b** data intervals, [1.88, 12.43] and [−4.94, 9.37]; $\Delta r = 1.78$

densities tend to take a multimodal shape, in the numerical experiments considered. Furthermore, the corresponding *bc*-based and *mb*-based random field estimates exhibited considerable differences in some cases. Finally, it was observed that in several numerical experiments (not shown here) the soft data incorporated at the meta-prior stage had a greater influence on the *mb*-based spatiotemporal estimates than on the *bc*-based estimates of the map (see also Example 5.9). ▪

6.4 Quantifying the Mapping Efficiency of Soft Data

When dealing with the potential use of soft datasets, it is convenient to define an *efficiency index* that quantifies the benefit of integrating this extra source of information into *TGIS* mapping (see also the theoretical discussion in Sect. 3.3.1). Although *BME* can use a wide variety of general knowledge sources, here we assume that the general knowledge consists of the mean and covariance (or variogram) functions. In this case, using the well-known fact that the Kriging variance is not influenced by the data values themselves (Olea 1999), one can define an efficiency index that characterizes the amount of additional information offered by the soft dataset. Two bounds can be assumed for such a purpose: One of them is the estimation variance when no soft data are used, and the other is the estimation variance when all soft data are replaced by hard data (both are Kriging variances). The *BME* estimation variance is expected to lie between these two bounds. As a consequence,

A useful efficiency index can be defined as the estimation variance reduction obtained using soft data divided by the difference of the corresponding Kriging variances.

This index varies from 0 to 1 and corresponds to the percentage of variance reduction brought by the soft data compared to what would have been obtained if, instead, only the hard data had been used. An efficiency index value close to 1 indicates that the soft data possess almost the same information content as the hard data, whereas an index close to 0 indicates that there is almost no information content offered by the soft data. Since this efficiency index may depend on the location of the estimation point relative to the data points, the index should be viewed as a local measure of the soft data efficiency. The use of the efficiency index is better illustrated with the help of a simple numerical example, as follows.

Example 6.13.[13] Realizations of a (0,1) Gaussian random field $X(s)$, $s = (s_1, s_2)$, are generated over a 20×20 regular grid covering a square area D of unit size, using the popular random field simulation technique implemented in the previous examples. The field is spatially homogeneous and its variogram has an exponential shape (its range and sill are both equal to 1). Assume that the field $X(s_1, s_2)$ is sampled at 50 locations as shown in Fig. 6.15a. At 25 of these locations, the sampled values are considered as hard data, whereas at the remaining 25 locations the simulated (hard) data are replaced by interval soft data of width equal to 1.5 units. Given that the field is

[13]The reader can use the routine ***example06.m*** to reproduce the results of this example.

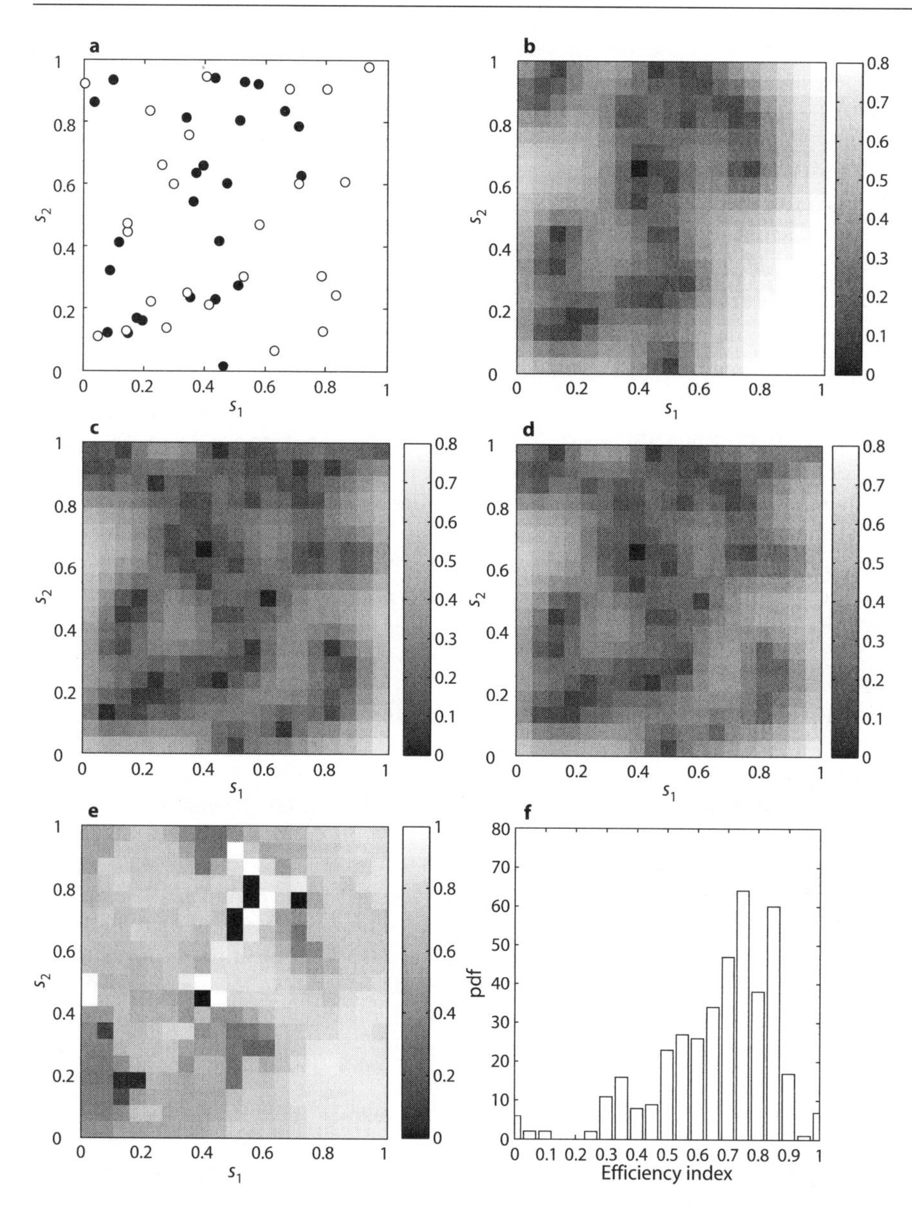

Fig. 6.15. a Locations of the 25 hard data (*black spots*) and the 25 interval data (*white circles*); **b**, **c** and **d** maps of estimation variances for Kriging with 25 hard data (*Kb*), Kriging with 50 hard data (*Ka*) and *BMEmean*, respectively. **e** Map of the efficiency index values. **f** Histogram of the efficiency index values in **e**

(0,1) Gaussian, these intervals are considered as weakly informative. Next we used three mapping techniques to produce a spatial estimation map covering D. The first technique considered was simple Kriging based on the 25 hard data (the technique is herein denoted as *Ka*, and the resulting estimation variance map is plotted in Fig. 6.15b). The estimation variance values range from zero (at the hard data points) to over 0.8 (away from the hard data points). A second simple Kriging technique (*Kb*) accounts for all 50 simulated hard data (Fig. 6.15c). Finally, the *BME* technique uses the 25 hard and the 25 soft (interval) data, thus generating the *BMEmean* estimation variance map of Fig. 6.15d. As Fig. 6.15e,f demonstrate, despite the fact that the width of the interval data considered is rather large, the estimation variance reduction obtained by *BME* is, on the average, close to 70% of the estimation variance reduction obtained by assuming hard data at the soft data locations (*Kb*). An obvious conclusion is that,

There is considerable benefit to using BME to incorporate these soft data in the mapping process.

As a matter of fact, the areas in which the estimation variance decreases clearly correspond to the areas where no information is available for the *Ka*-based mapping. Moreover, the reader should notice the close similarity between the estimation variance maps obtained by the *Kb* technique (Fig. 6.15c) and the corresponding maps obtained by the *BME* technique (Fig. 6.15d). The *BME*, though, uses only half the number of hard data that the *Kb* uses and, thus, it is a much less expensive mapping technique to apply in *TGIS* practice. ▪

6.5 Numerical Investigations of Popular Techniques

6.5.1 The Use and Misuse of Soft Data by Statistical Regression-Based Techniques

Although statistical regression-based techniques (e.g., objective analysis and Kriging) are rather popular for processing hard datasets across space and time, there are certain unresolved problems when processing soft information. Some of these problems were discussed in the preceding sections. For additional illustration, we focus in this section on interval soft data, using numerical experiments to investigate the results obtained from different mapping techniques, and we emphasize the importance of assessing correctly the measurement uncertainty.

Assume that a *TGIS* specialist decides to employ a space/time regression technique to process data using some kind of a "trick". One such commonly used trick is the so-called *hardening* of the soft data, which consists of replacing each interval (soft) datum by a single hard value, say, the middle value of the interval (e.g., Voltz et al. 1997). One then needs to examine the effect of hardening on the resulting maps. Intuitively, hardening should be expected to yield similar estimates as *BME* if the widths of the interval data are very small compared to the range of possible field values. As the interval becomes wider, one would expect that the soft data will have a smaller effect on space/time estimation, since these intervals become less informative. Unfortunately, many of the commonly used tricks do not account for such an intuition. If we assume, e.g., that an interval datum is broadened arbitrarily but without changing the middle value used in the hardening trick, there will be no change in the space/time regression estimates, which is a result that does not make any sense. An expert could argue that beyond a specific width the interval datum becomes useless and should not be considered in *TGIS* mapping, but problems arise when one has to define this width. Such limitations of space/time regression techniques can be overcome by using the *BME* approach, since the latter is specifically developed to take soft data rigorously into account and does not involve any questionable tricks. The effectiveness of *BME* and, at the same time, the inadequacy of space/time regression in such cases will be demonstrated with the help of the following examples. For the purpose of comparing *BME* with other techniques (space/time regression, Kriging, etc.) it is assumed that the general knowledge consists of the mean and covariance (or variogram) functions.

Example 6.14.[14] Consider a (0, 1) Gaussian random field $X(\boldsymbol{s})$, $\boldsymbol{s} = (s_1, s_2)$, which is spatially homogeneous and has an exponential variogram (its range and sill are both equal to 1). Simulated samples of the field are available at 125 locations over a square area D of unit size (Fig. 6.17a). At 25 of these locations the values obtained are considered as hard data. For the remaining 100 locations, the only case-specific knowledge assumed available is that the corresponding values belong to certain known intervals. Three intervals are used, as follows: [−3.09, −0.67], [−0.67, 0.0], and [0.0, 3.09]. In Fig. 6.16 these intervals are defined in terms of the limits of classes established in terms of the 0.001, 0.25, 0.5 and 0.999 quantiles of the (0, 1) Gaussian distribution. For comparison purposes, values of the actual field X are simulated at the 400 nodes of a 20×20 grid covering D (Fig. 6.17b). The objective is to compare the maps of estimates produced at the 400 nodes by using

i. Kriging based on hard data only (denoted as *KH*).
ii. Kriging based on hard data and "hardened" soft data (*KHS*).
iii. The *BME* technique based on hard and soft data.

For the purpose of the present analysis, in the case of the *KHS* technique the 100 soft data were "hardened" by considering the middle of each interval as a hard datum

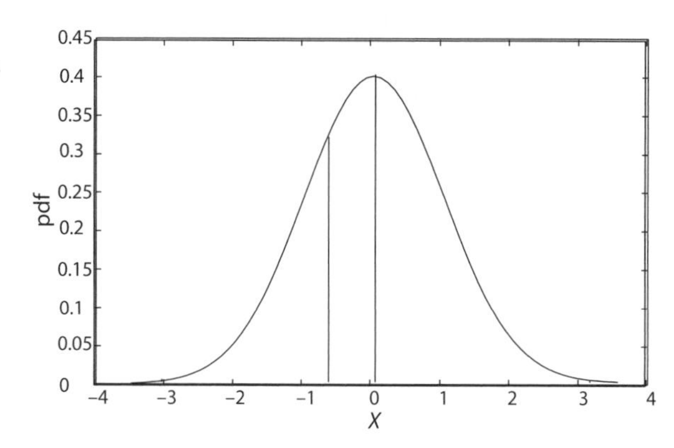

Fig. 6.16. Definition of soft data intervals in terms of 0.001, 0.25, 0.5, and 0.999 quantiles of the (0, 1) Gaussian distribution

[14]The reader can use the routine ***example07.m*** to reproduce the results of this example.

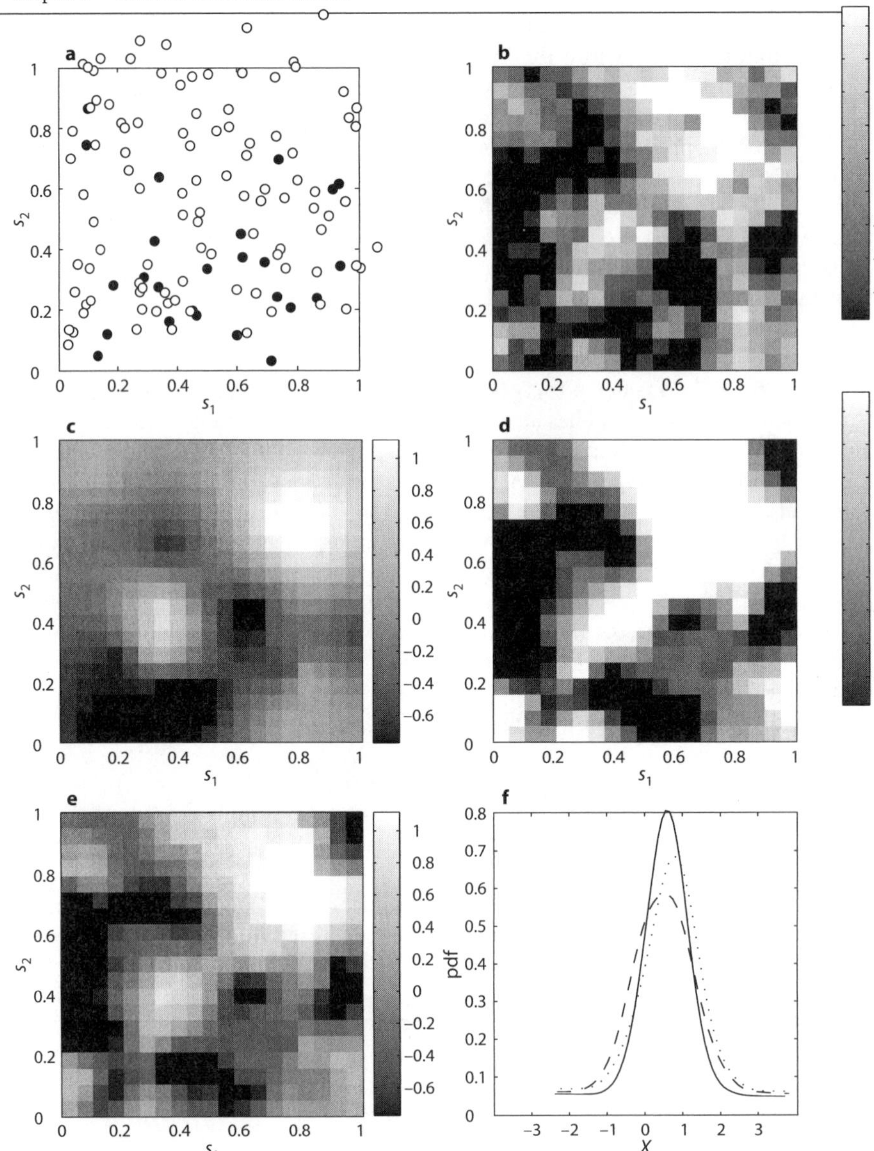

Fig. 6.17. **a** Locations of the 25 hard data (*black spots*) and the 100 soft data (*white circles*); **b** simulated map; **c** *KH* map, **d** *KHS* map, and **e** *BMEmode* map; **f** error distributions for *KH* (*dotted line*), *KHS* (*dashed line*), and *BME* (*plain line*)

(clearly, there were only 3 possible values, i.e., −1.88, −0.34 and 1.55). Unlike the *KHS* technique, *BME* does not need to employ such tricks, since its rigorous mathematical structure allows it to incorporate directly the corresponding interval (soft) data. The estimated maps (*KH*, *KHS*, and *BMEmode*) are plotted in Fig. 6.17. These maps demonstrate convincingly the need to account adequately for soft data. In particular, *KH*

Table 6.6.
Estimation errors

	Mean error	Mean square error
KH	−0.11	0.39
KHS	−0.23	0.51
BME	−0.07	0.21

yields a map (Fig. 6.17c) that is much too smooth and does not adequately reproduce the actual variability of the simulated values (Fig. 6.17b). When the additional information at the 100 locations is included, *BME* (Fig. 6.17e) gives better results than both *KH* and *KHS* (Fig. 6.17d). The *KHS* map exhibits a bias toward higher values (note the white area that occupies the center of the map). This bias is easily explained by the asymmetry of the intervals considered, whereas the poorer performance of *KHS* vs. *KH* is linked to the fact that the Kriging algorithm has been deceived (so to speak) by the hardening trick. Indeed, *KHS* gives too much weight to the hardened soft data by processing them like they were hard data. This is especially valid for the last interval $[0.0, 3.09]$, which is the least informative. From an estimation error viewpoint (see Fig. 6.17f), *KHS* performs very poorly even compared to *KH*. The *BME* technique, on the other hand, yields a map with the smallest estimation errors (see also Table 6.6). The *BME* error distribution (Fig. 6.17f) is centered around zero and exhibits a smaller variability than the *KHS* error distribution. This is because, unlike *KHS*,

BME successfully distinguishes between the different information contents of the soft data.

Despite the fact that the three intervals may be considered as not particularly informative (the probabilities that the intervals contain a datum are, 25%, 25%, and 50%), they still lead to serious mapping improvements when used rigorously by the *BME* technique. ▪

Some other attractive features of the *BME* method are illustrated in the following numerical experiment, which assumes that soft data are available throughout the entire area of interest.

Example 6.15.[15] In Fig. 6.18 we consider the same arrangement as in Example 6.14, but now assume that only soft data of the interval type are available for the field $X(\boldsymbol{s})$, $\boldsymbol{s} = (s_1, s_2)$, at 100 locations over the area D (Fig. 6.18a). The results obtained in this case (Fig. 6.18b–d) show that the map of the *BMEmode* estimates is considerably better than the map of the *KHS* estimates (see also the associated error distributions in Fig. 6.18e). Moreover, the *KHS* estimation errors are even worse than the corresponding errors in Example 6.14. Again, a *KHS* bias exists towards higher values (which is rather easily explained by the asymmetry of the interval data). The *BME* method yields a map which exhibits considerably smaller estimation errors than the *KHS*-based map (see the estimation errors in Table 6.7). ▪

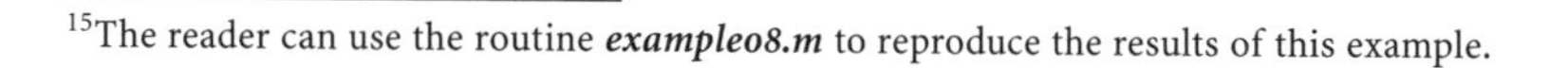

[15]The reader can use the routine ***example08.m*** to reproduce the results of this example.

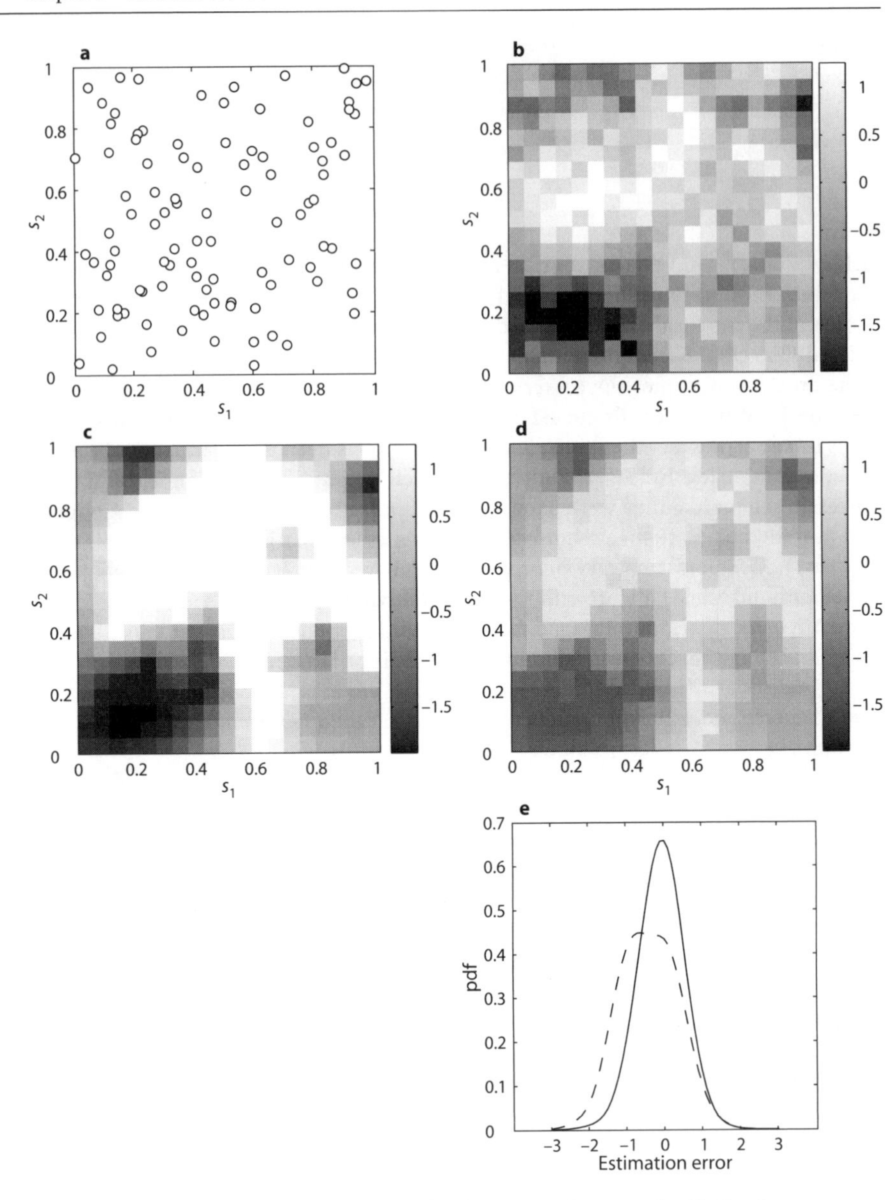

Fig. 6.18. **a** Locations of 100 soft data; **b** simulated map; **c** *KHS* map, and **d** *BMEmode* map; **e** error distributions for *KHS* (*dashed line*) and *BME* (*plain line*)

Table 6.7.
Estimation errors

	Mean error	Mean square error
KHS	–0.42	0.69
BME	–0.09	0.29

As is shown in the following example, the use of the so-called "hardened" soft data by some mapping techniques can lead to absurd situations, in which no meaningful estimates can possibly be derived. In the case of the Kriging techniques, e.g., these unpleasant situations are the result of some of its inherent properties (e.g., exact interpolator) which, while providing useful estimates when only hard data are considered, may lead to some serious problems when Kriging is requested to incorporate soft data into the mapping process.

Example 6.16.[16] The domain of the field $X(s)$, $s = (s_1, s_2)$, is the same as in the previous examples. Consider 50 hard data points irregularly located over the square area D (Fig. 6.19a), and 400 interval soft data located at the mapping nodes of the usual two-dimensional grid. The procedures for simulating hard and soft data are the same as in the previous two examples. The resulting maps (Fig. 6.19c–e) show that both the *KH* and *KHS* techniques are clearly inadequate. In particular,

The KH technique uses hard data only (neglecting any additional soft data), whereas the KHS technique makes a very poor use of the soft data and undervalues the contribution of the hard data.

The reason for this poor performance is that the *KHS* technique is an exact interpolator and, thus, it honors the data values available at the estimation points (which, in this case, are the "hardened" soft data; i.e., –1.88, –0.34 and 1.55). But, the *KHS* technique completely neglects any data (hard and/or soft) at neighboring points, hence yielding merely a spatial map of the middle values of the interval data at the mapping nodes. The resulting map is, of course, rather useless. Moreover, the reader may notice that the *KH* technique performs better than the *KHS* technique (see estimation errors in Table 6.8). On the other hand, the *BME* approach leads to satisfactory results, as is easily seen in the spatial *BMEmode* map of Fig. 6.19e and in Table 6.8. This map accurately reproduces the main features of the simulated map (shown in Fig. 6.19b). Also, the *BME* map seems to integrate some meaningful features appearing separately in the *KH* and the *KHS* maps (e.g., global vs. local variations), thus implying that the different information sources used by the *KH* and the *KHS* techniques have been merged in a single map by the considerably more general *BME* method. ■

In situations in which the widths of the interval (soft) data used in *TGIS* mapping are large compared to the variation of the underlying spatial and/or temporal distribution, little weight is given to the soft data. Naturally, the soft data become more informative when the widths of the intervals are reduced. This situation can be illustrated with the help of two extreme cases:

- When intervals of infinite widths are considered, the soft data are ignored by *BME*, which generates a map that coincides with the *KH* map obtained using only hard data.
- When the width of the interval is equal to zero, the soft data become hard data and the *BMEmode* map coincides with the *KH* map obtained by including these additional hard data[17].

[16] The reader can use the routine ***example09.m*** to reproduce the results of this example.
[17] These practical results can be proven mathematically on the basis of the *BME* equations; Christakos (2000).

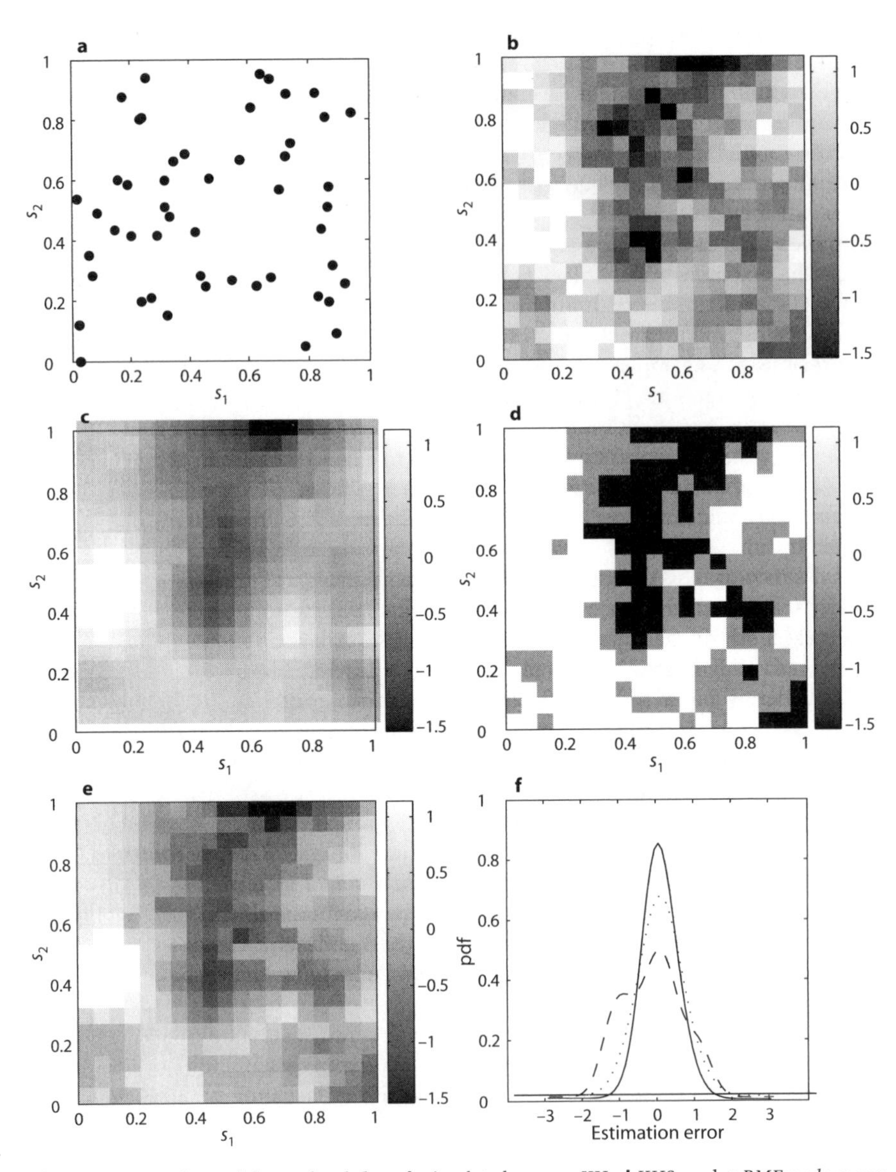

Fig. 6.19. a Locations of the 50 hard data; **b** simulated map; **c** *KH*, **d** *KHS*, and **e** *BMEmode* maps; **f** error distributions of *KH* (*dotted line*), *KHS* (*dashed line*) and *BME* (*plain line*)

Table 6.8.
Estimation errors

	Mean square error
KH	0.28
KHS	0.61
BME	0.14

Hence, these two extreme situations are properly handled by the *BME* approach which provides a physically consistent assessment of the quality of the soft information. The same, however, is not true for the *KHS* method. Instead, the larger the width of the intervals is, the more dubious the results obtained by *KHS*. In the following example, it will be shown with the help of numerical illustrations how the *BME* approach is able to account for the varying quality of the available soft data adequately.

Example 6.17.[18] A set of 25 hard data, 100 soft data, as well as 400 values at the usual grid nodes over the square area D are simulated for the spatially homogeneous random field $X(s)$, $s = (s_1, s_2)$, using the simulation technique of the previous examples (Fig. 6.21a,b). Two different definitions of interval data are considered:

a In the first definition, the limits of the corresponding 10 interval data classes are the deciles of the (0,1) Gaussian distribution (Fig. 6.20a).
b In the second definition, there are only 2 intervals determined by the limits of the interval classes which are the 0.001, 0.5, and 0.999 quantiles of the Gaussian distribution (Fig. 6.20b), i.e., the intervals $[-3.09, 0]$ and $[0, 3.09]$.

As in the previous example, spatial estimates are generated at the 400 grid nodes using the *KH*, *KHS*, and *BME* techniques. The first *KH* map (Fig. 6.21c) is obtained using 125 simulated hard values, whereas the second *KH* map (Fig. 6.21d) uses only the 25 simulated hard values. Clearly, the *KHS* maps (Fig. 6.21e,f) and the *BME* maps (Fig. 6.21g,h) are almost identical with the *KH* maps in the case (*a*), thus showing that the information content of the soft dataset is close to the information content pro-

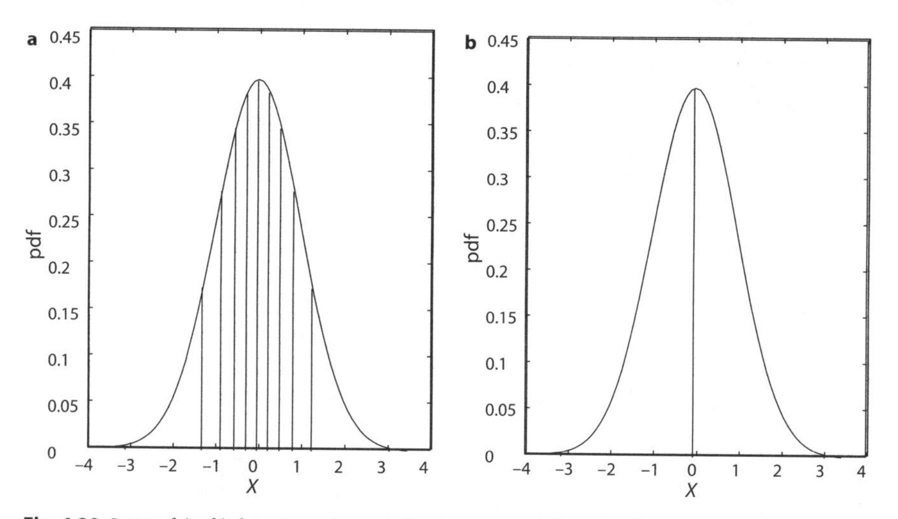

Fig. 6.20. Interval (soft) data. Bounds are defined in terms of the: **a** deciles of the (0,1) Gaussian pdf; **b** the 0.001, 0.5 and 0.999 quantiles of the (0,1) Gaussian pdf

[18]The reader can use the routine ***example10.m*** to reproduce the results of this example.

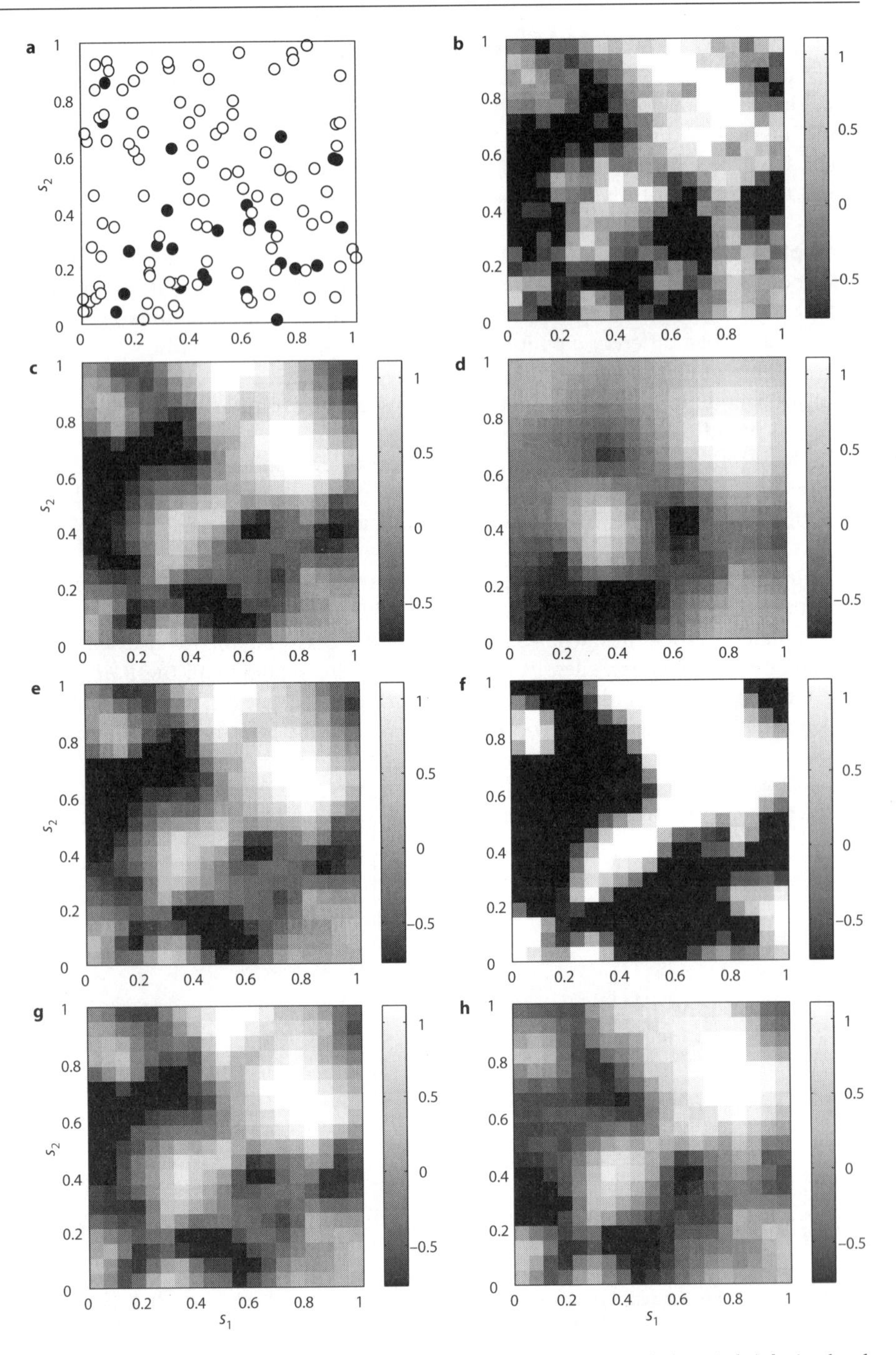

Fig. 6.21. a Locations of the 25 hard data (*black spots*) and the 100 soft data (*white circles*); **b** simulated map; **c** *KH*, **e** *KHS* and **g** *BMEmode* maps according to the interval data definition of Fig. 6.20a; **d** *KH*, **f** *KHS* and **h** *BMEmode* maps according to the interval datum definition of Fig. 6.20b

Table 6.9.
Estimation errors

	Mean squared error	
	10 classes	2 classes
KH	0.19	0.38
KHS	0.21	0.66
BME	0.20	0.24

vided by the simulated hard data at the same points. As can be seen in Table 6.9, the estimation errors obtained by all three mapping techniques in case (*a*) are very close to each other. However, the maps produced by *KH*, *KHS*, and *BME* differ considerably in the case (*b*) above. Under these conditions, by giving too much weight to the middle of the interval datum, the *KHS* technique produces a map (Fig. 6.21f) that exhibits a very contrasted pattern (it looks almost like a binary map). Again, instead of improving spatial estimation, the "hardened" soft data used by *KHS* lead to a poor performance in terms of estimation errors. Nevertheless, the fact that there is still valuable information in the soft dataset is reflected in the better performance of *BME* mapping compared to *KH* mapping (see estimation errors in Table 6.9). ▪

The preceding Examples 6.14–6.17 demonstrated some significant advantages of the *BME* approach over some popular mapping techniques. For instance,

While other popular techniques performed poorly when they used ad hoc *tricks to account for soft datasets, BME made a clear distinction between hard and soft data, adequately accounted for the different information contents of the data sources, and led to accurate and informative maps.*

This kind of *BME* traits can be used profitably in a variety of *TGIS* applications in the life support sciences.

6.5.2 The Inadequacy of Indicator Kriging

Indicator Kriging (*IK*; see also Sect. 4.3) is a geostatistical technique that allows the extraction of some information from imprecise data, although not always in a mathematically rigorous and physically meaningful fashion. The *IK* technique is based on a data coding scheme which should be able, in theory, to account for both hard and soft datasets simultaneously. For a given threshold, a hard datum is coded as either 0 or 1, depending on whether the value is above or below the threshold; a soft datum is coded using the associated cdf values for that threshold (Deutsch and Journel 1992). Despite its apparent simplicity, *IK* suffers from certain severe limitations. For a detailed discussion of the theoretical and practical limitations of *IK*, see e.g., Olea (1999), and Chiles and Delfiner (1999). Below we give a brief review of some of these limitations:

- Cdf estimates obtained from *IK* do not respect some fundamental conditions, like the requirement that the probability values should lie within the [0, 1] interval, or that the cdf should be a monotonic function (as a non-convex estimator, *IK* can not easily incorporate this kind of condition).

- Estimating cdf for a given threshold requires knowledge of the corresponding indicator variogram for that threshold, which leads to serious modelling and estimation problems. For example, there are as many indicator variograms to be modelled and *IK* systems to be solved as the number of thresholds considered.
- Additional problems arise when using the indicator *co-Kriging* technique (considered by some geostatisticians as the most complete indicator approach), including a huge increase in computation time. It is customary, e.g., to model the entire set of (cross-)variograms using the same set of basic variogram models, but this approach does not guarantee the validity of the bivariate cdf and can lead to very crude approximations.

Although *IK* is often presented as a non-linear method, the truth is that

The only non-linear element of IK is the coding of the data; data processing is still done using linear Kriging algorithms. On the other hand, BME is a fully non-linear technique.

It is also worth noticing that replacing the exact data values with the indicator coding results to a loss of information that can be serious if the coding is coarse (e.g., when only a few thresholds are defined). In some cases, this loss may be reduced by using a finer coding resolution, but this will increase considerably the computation time. Taking a clue from the above remarks, some of the *IK* limitations are illustrated below by means of a synthetic example. The performance of classical Kriging with hard data (*KH*), *BME*, and *IK* are compared. The numerical results show that even under the multivariate Gaussian assumption the *IK* performs poorly, whereas *BME* performs in a very satisfactory manner.

Example 6.18.[19] Values of a (0, 1) Gaussian random field $X(\boldsymbol{s})$, $\boldsymbol{s} = (s_1, s_2)$, with an exponential covariance (its range and variance are both equal to 1) are simulated over a square area D, using the simulation technique of the previous examples. The objective of this study is to compare the various cdf obtained in terms of *KH*, *IK*, and *BME*. Consider 50 arbitrarily sampled locations over D. The *KH* technique is allowed to use the simulated (actual) field values at the 50 locations as hard data, whereas in the case of the *BME* and *IK* techniques the 50 simulated values will be replaced by certain intervals to which they belong. In particular, the limits of these interval data are the $0.001, 0.1, 0.2, \ldots, 0.9$ and 0.999 quantiles of the (0, 1) Gaussian distribution (Fig. 6.22a). This arrangement gives *KH* a considerable advantage and allows the *KH*-based plots to serve as reasonable references in some comparison cases. Estimates of the random field using the above techniques are obtained on a 20×20 grid covering D. Since *IK* requires knowledge of the indicator covariances (or variograms) for every threshold, the corresponding theoretical covariances are plotted in Fig. 6.22b. The different shapes of the ordinary vs. the indicator covariance models may be explained by the fact that,

[19]The reader can use the routine ***example11.m*** to reproduce the results of this example.

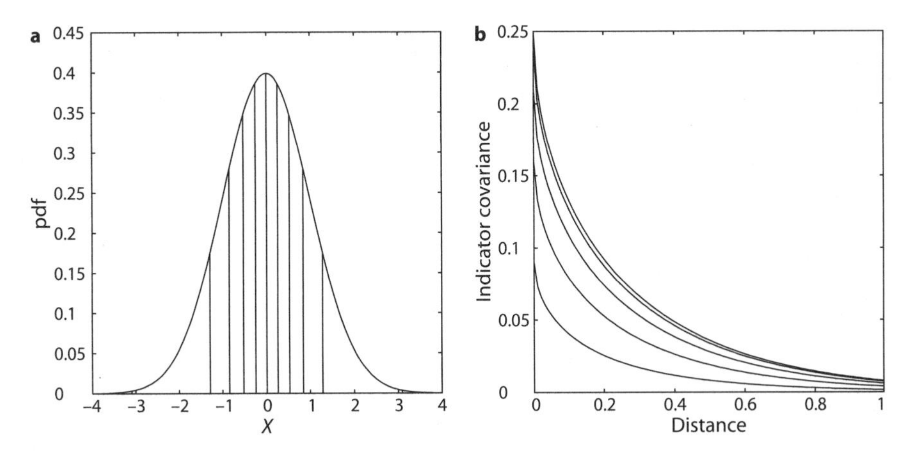

Fig. 6.22. Indicator data coding. **a** Class limits defined as the 0.001, 0.1, 0.2, ..., 0.9 and 0.999 quantiles of the (0, 1) Gaussian distribution; **b** indicator covariances for the 0.001, 0.1, 0.2, ..., 0.5 quantiles (covariances are symmetric around the 0.5 quantile)

the theoretical indicator covariances (Fig. 6.22b) are obtained by integrating with respect to the bivariate Gaussian distribution, and they do not necessarily have the shape of any of the traditional covariance functions (exponential, Gaussian, etc.)[20]. For spatial mapping purposes, the *KH* and *BME* techniques use the ordinary covariance (exponential model), whereas the *IK* technique uses the indicator covariance models of Fig. 6.22b. This allows a reasonable comparison of the relative performances of these mapping techniques, without having to consider complicated estimation problems and methodological questions. In a sense, the estimation results under these circumstances should represent the best results that one could expect from these three mapping techniques. The cdf calculation in the case of *KH* uses the fact that under the multivariate Gaussian assumption the cdf at the mapping nodes are also Gaussian and are characterized by their mean and variance (i.e., the simple Kriging estimate and its error variance). In the case of the *BME* technique the cdf calculation is an easy affair for any arbitrary threshold value, whereas for the *IK* technique the values of the cdf can be calculated only for the threshold values previously defined (due to the fact that different indicator covariances must be used for these thresholds). Figure 6.23 amply demonstrates that, despite the loss of information due to hard data coding into intervals, the cdf plots obtained by *BME* are very close to those obtained by *KH* (which may serve as the reference plots, in this case). On the other hand, the *IK* technique produces very poor results compared to the reference plots. Indeed, it is worth noticing that,

About 35% of the cdf calculated by IK do not respect the monotonicity condition, and 5% of them yield probability values outside the [0, 1] interval.

[20]This can be a problem in some cases, given that it is a common geostatistical practice to fit these traditional functions to raw indicator covariances.

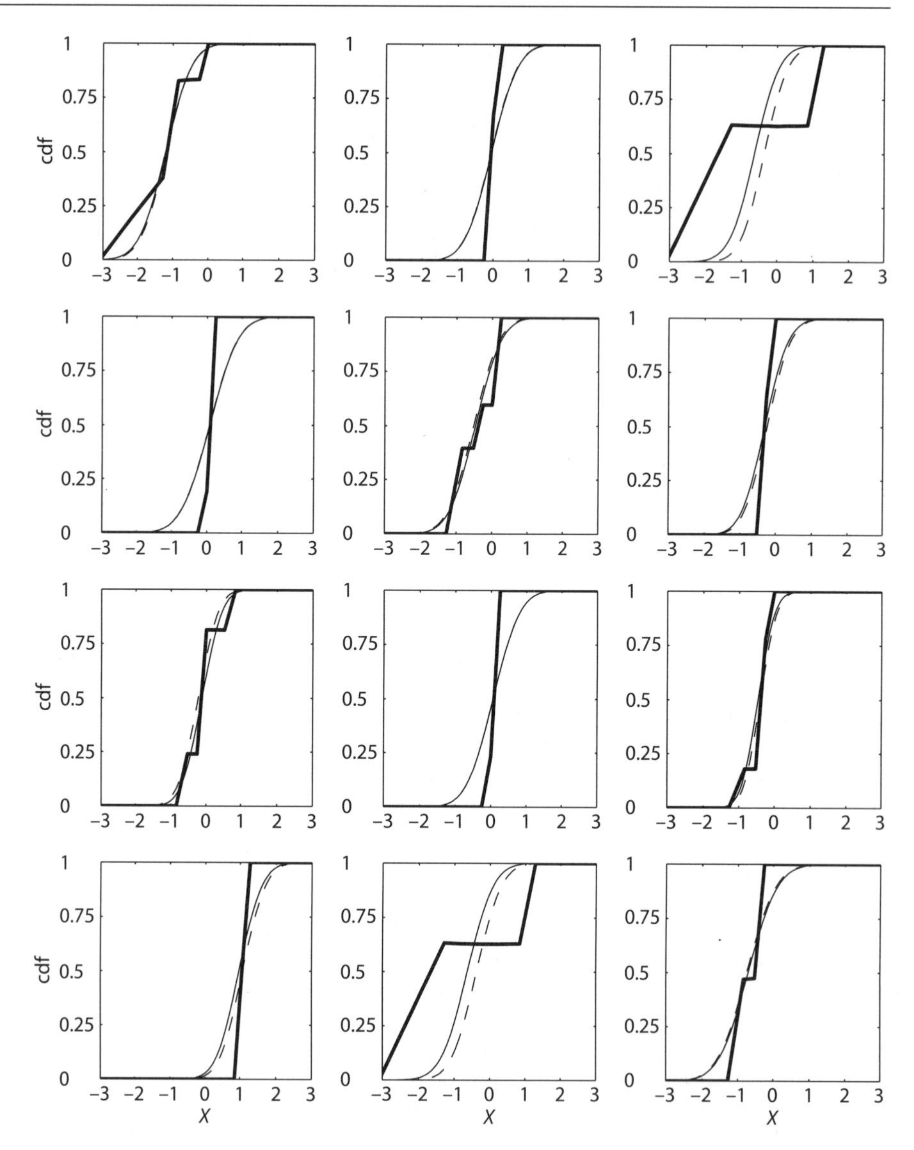

Fig. 6.23. Cdf calculated at 12 randomly selected grid nodes (from a total of 400 nodes). *Dashed*, *plain*, and *bold lines* correspond to *KH*, *BME* and *IK* estimates, respectively (some of the cdf plots for *BME* and *KH* are so close that they cannot be easily distinguished)

The poor performance of *IK* can be, also, seen with the help of maps depicting the probability that the estimated field values (obtained from *KH*, *BME*, and *IK* at specified locations) belong to the specified interval [−0.52, 0.52]. As is shown in Fig. 6.24 both *KH* and *BME* yield identical results, whereas *IK* yields a much more contrasted map with plenty of values close to 0 or to 1. Furthermore, when the *TGIS* specialist needs to decide whether the field value at a given location belongs or not to a speci-

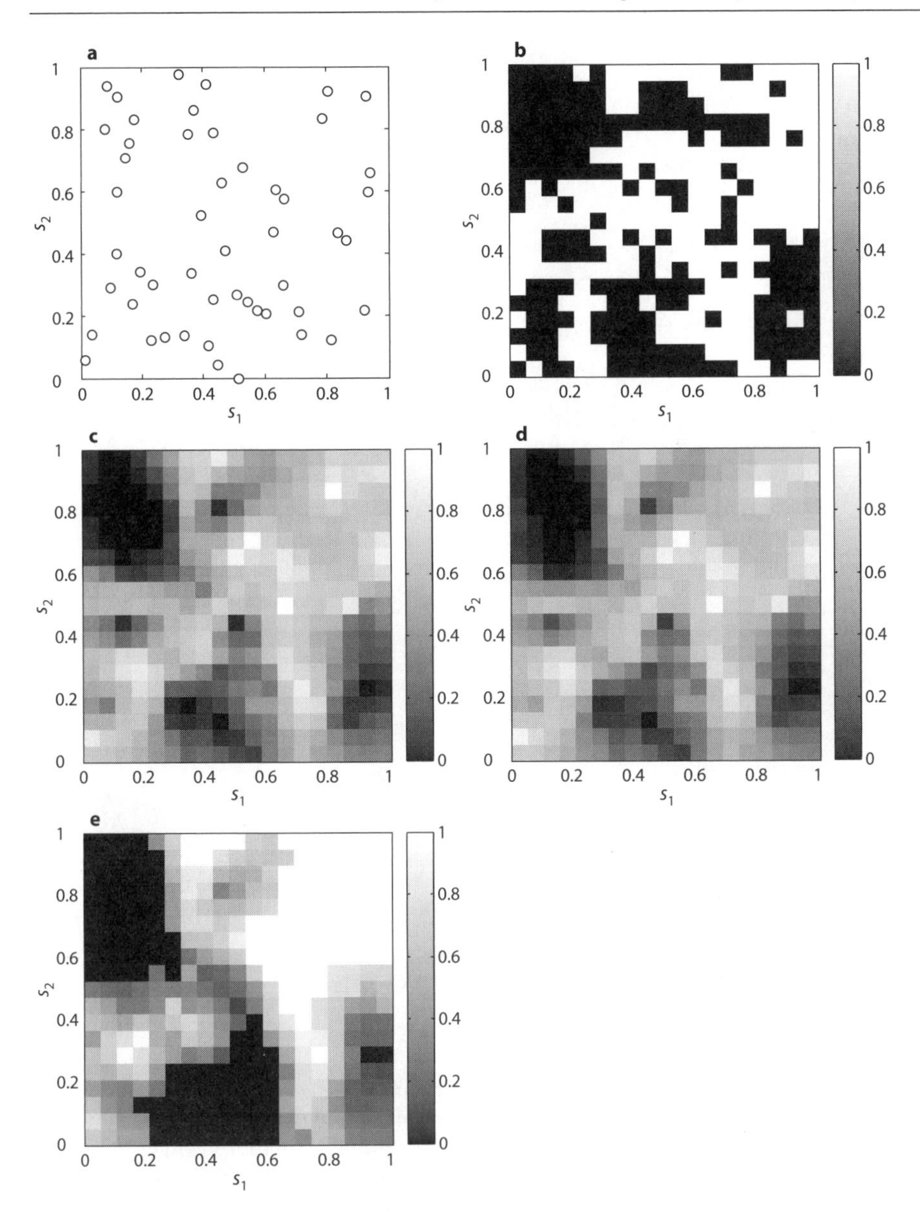

Fig. 6.24. a Locations of 50 hard data; **b** simulated map (shown in white are data belonging to the [−0.52, 0.52] interval); **c, d** and **e** probability maps of belonging to this interval obtained by *KH*, *BME*, and *IK*, respectively

fied interval, he is faced with the choice of a probability threshold (i.e., in many cases the specialist has to select a threshold value such that, if the predicted probability of belonging to the interval exceeds this value, he will accept the hypothesis that the value actually belongs to the interval; otherwise he will reject the hypothesis). The *BME* tech-

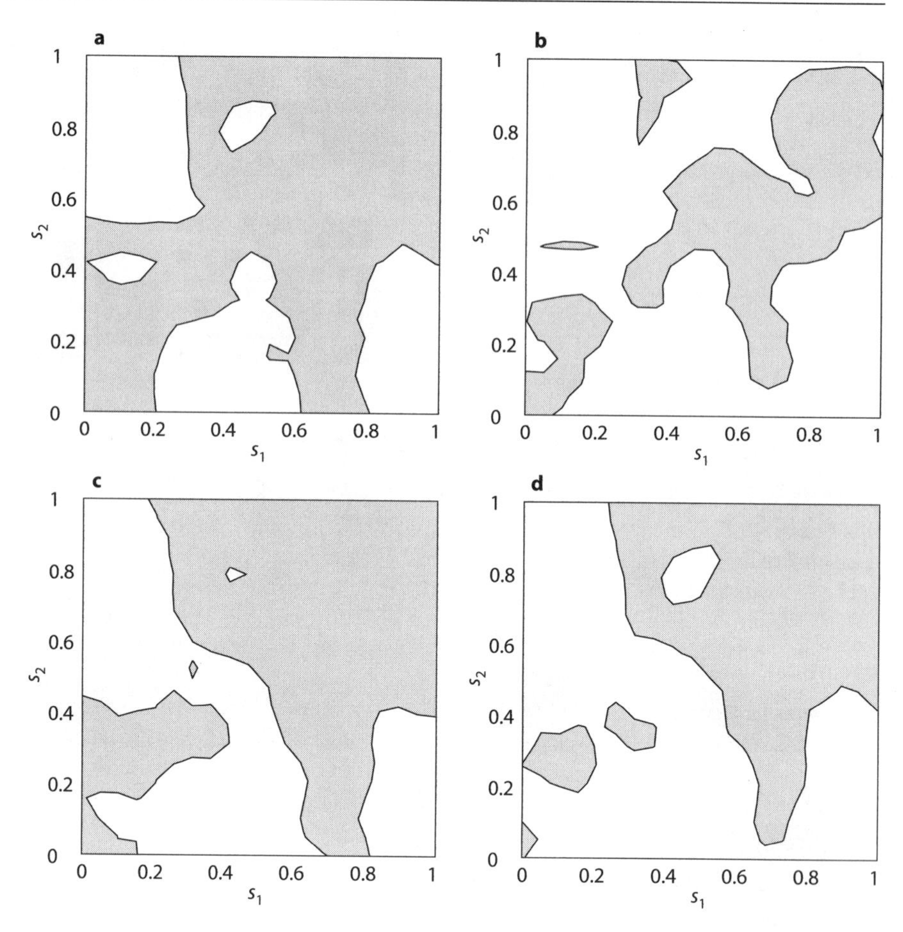

Fig. 6.25. *Left* and *right columns* show the areas that belong to the interval [−0.52, 0.52] (*shaded areas*) for the 0.4 and 0.6 probability thresholds, respectively. Upper and bottom rows show the maps for *BME* and *IK*, respectively

nique may lead to different outcomes when this probability threshold is changed, e.g., from 0.4 to 0.6 (see Fig. 6.25). Also, while the *BME*-based decision map exhibits considerable modifications, the *IK*-based maps show much smaller changes. It can be shown that the correct classification rate of the *IK* technique (i.e., the % of values that have been correctly selected as belonging or not to an interval) is rather insensitive to the value of the probability threshold, whereas there is an optimum threshold value in the case of the *BME* and *KH* technique. The apparent robustness of the *IK* results with respect to the selection of the probability threshold may cause some problems. As a matter of fact, controlling this probability threshold is very important in many applications in which the user wants to be protected from potentially disastrous consequences (e.g., when the risk of error has a serious financial, health, environmental, ecological etc. impacts, the decision-maker will tend to select high probability thresholds, in order to reduce the risk of wrongly deciding that the value at a specified loca-

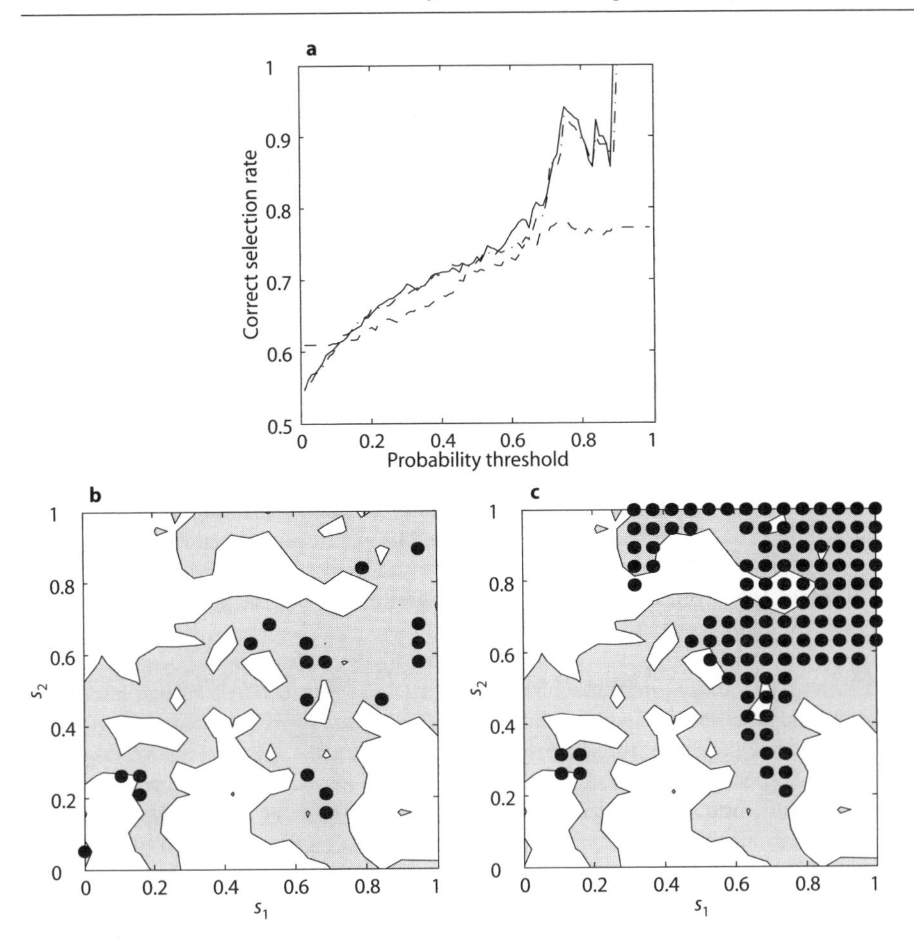

Fig. 6.26. a Rate of correctly selected values vs. probability threshold (*dashed-dotted, plain*, and *dashed lines* correspond to *KH*, *BME*, and *IK*, respectively); **b** and **c** grid nodes selected by *BME* and *IK*, respectively, to belong to interval [−0.52, 0.52] for a probability threshold of 0.8 (*shaded areas* correspond to simulated values which actually belong to this interval)

tion belongs to an interval). This is almost impossible to achieve in terms of the *IK* technique. On the basis of the numerical experiments of this study, one concludes that even when a probability as high as 0.8 is used, about 25% of the locations are wrongly selected by *IK* to respect this condition, compared to only about 10% by *BME* (see Fig. 6.26a). Furthermore, the maps of Fig. 6.26b,c show that the *IK* technique seems to constantly select locations at the upper-right part of the map as belonging (in several cases incorrectly) to the interval [−0.52, 0.52]. The *BME* technique, on the other hand, is much more cautious, selecting fewer locations throughout the map as belonging to the same interval (in all cases except one the selections are correct). By way of a summary, by assuming a high probability threshold, the *TGIS* specialist may assume that he is on the safe side with the *IK* technique, when in reality this is not true. Only the *BME* technique is doing an adequate job, in this case. ▪

Furthermore, numerical studies of categorical variables have shown that the *BME* technique performs much better than the *IK* or indicator co-Kriging (*Ico-K*) techniques when used to estimate the probabilities of belonging to specified categories in space and time (Bogaert 2001).

6.6 Merging Maps with *BME*

It is not uncommon in *TGIS* applications that several maps are available representing essentially the same physical or ecological field. These maps could be coming from different information sources or spatiotemporal analysis techniques (e.g., the maps may be the outputs of different numerical models or in situ sampling campaigns). Since all the above maps refer to a specific aspect of physical reality and, perhaps, they are reliable to various extents, the *TGIS* specialist is faced with the difficult decision of selecting the most appropriate map for the situation under consideration, deciding which characteristics of each map are valuable for the purposes of the study and which are not, etc. The *BME* approach is a useful tool in this respect, since

BME allows map-merging in a manner that properly incorporates the relevant information amounts contained in the various maps.

This map-merging property of *BME* is illustrated with the help of two numerical examples, as follows.

Example 6.19.[21] In Fig. 6.27a we consider a (0,1) Gaussian pdf, which plays the role of the prior pdf (i.e., this pdf was constructed on the basis of the general knowledge available about the natural field of interest). In the same figure we also plot a [−1, 6] uniform pdf which plays the role of the site-specific pdf at the same point in the space/time domain of the field[22]. For the purposes of the present analysis, one important piece of information provided by the uniform pdf is the fact that the corresponding field value cannot be less than −1. The posterior pdf obtained by the *BME* approach turns out to have a truncated Gaussian form, which is also plotted in Fig. 6.27a. As should be expected, the *BME*-based posterior pdf has successfully taken into account the significant pieces of information contained in the previous pdf.

In a separate case (see Fig. 6.27b) both the general knowledge-based pdf and the site-specific knowledge-based pdf are assumed to have Gaussian shapes (although, with different means and variances). Under these circumstances, an important piece of information is clearly the Gaussian shape of the two pdf. Indeed, it turns out that the *BME*-based posterior pdf has a Gaussian shape, as well (Fig. 6.27b). Furthermore, the reader may find it worth noticing that the posterior pdf has a smaller variance than any of the two component pdf, as well as a mean value that lies between the mean values of the two component pdf. ■

[21] The reader can use the routine ***example12.m*** to reproduce the results of this example.

[22] It is a classical result that a new pdf can be derived by combining two different pdf at the same spatial location by means of the *BME* formalism.

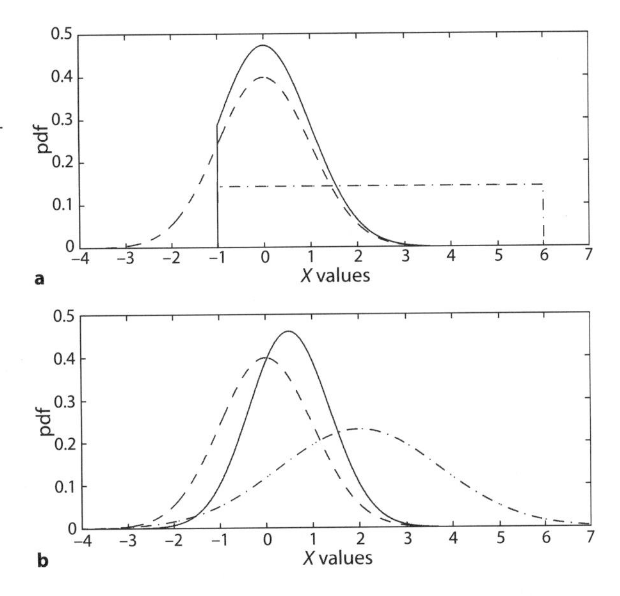

Fig. 6.27.
a (0,1) Gaussian and [−1, 6] uniform pdf; **b** (0,1) and (2,3) Gaussian distributions. *Plain lines* represent the corresponding *BME* pdf

Example 6.20.[23] In Fig. 6.28 values of a (0,1) Gaussian random field with an exponential variogram model (its range and sill are both equal to 1) are simulated throughout a unit square area D using the simulation technique of the previous examples (Fig. 6.28b). Consider 50 arbitrarily sampled locations over the square area (Fig. 6.28a). The simulated field values at these locations are considered hard data (let us denote this dataset by S). For the purpose of this study we divide the hard dataset S into two subsets. The first one, S_1, contains only field values that are less than zero, whereas the second one, S_2, contains values that are greater than or equal to zero (clearly, $S = S_1 \cup S_2$). Two simple Kriging maps (*KH*1 and *KH*2) are constructed based on the corresponding set of hard data (i.e., *KH*1 uses the subset S_1, whereas *KH*2 uses the subset S_2). The estimation results (see Fig. 6.28c,d) show that, while the *KH*1 and *KH*2 maps are adequate in areas where the corresponding subset of data points is dense, they are very poor in the other areas. The *BME*(1,2) map (Fig. 6.28f) is obtained by merging the previously obtained *KH*1 and *KH*2 maps. This is done on a point basis, i.e., at each mapping point a new pdf is derived that combines the two Gaussian pdf obtained separately with the help of the *KH*1 and *KH*2 techniques[24]. Indeed, as these two maps contain complementary information, a much more satisfactory *BME*(1,2) map is obtained by merging the two maps together. The biases of the *KH*1 and *KH*2 maps disappear in the *BME*(1,2) map, and the estimation error is considerably reduced. For comparative purposes, the *BME*-based map that was obtained by using the complete hard dataset S is also plotted in Fig. 6.28e. Overall, the *BME*(1,2) estimation errors are not much higher than the *BME* errors (see Fig. 6.28g and Table 6.10). Note that a theoretical difference between the two techniques, i.e. the original *BME* technique and the merging *BME*(1,2) technique, is that the latter rather neglects spatial correlations between the two separate datasets S_1 and S_2. ▪

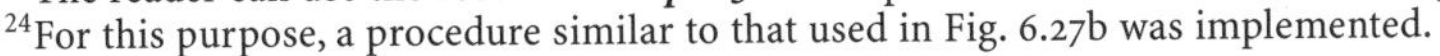

[23]The reader can use the routine ***example13.m*** to reproduce the results of this example.
[24]For this purpose, a procedure similar to that used in Fig. 6.27b was implemented.

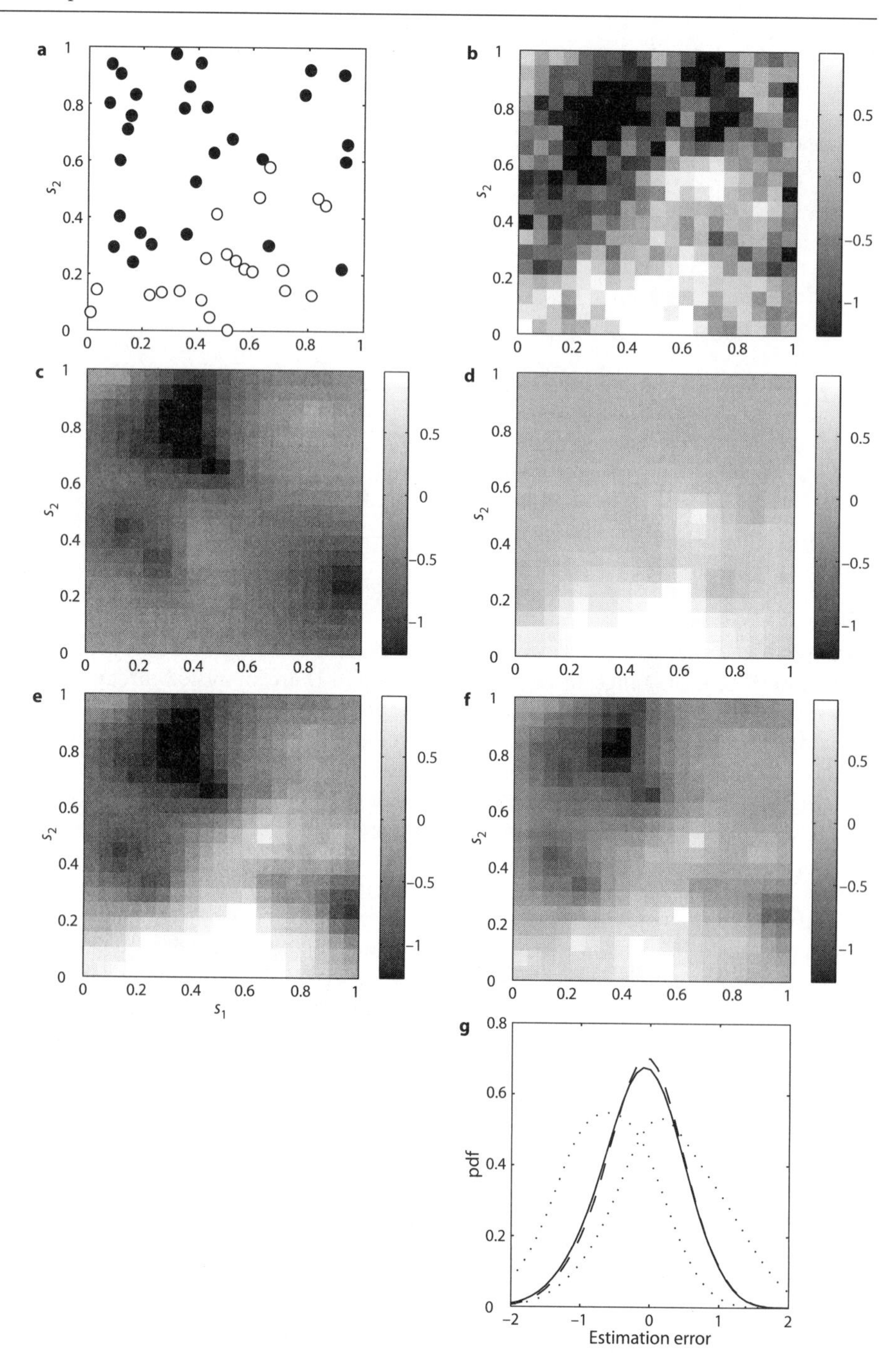

Fig. 6.28. a Locations of the hard data that are below zero (*black spots*) and equal or above zero (*white circles*); **b** simulated map; **c** *KH*1, **d** *KH*2, **e** *BME*, and **f** *BME*(1, 2) maps; **g** error distribution for *KH*1 and *KH*2 (*dotted lines*), *BME* (*dashed line*), and *BME*(1, 2) (*plain line*)

Table 6.10.
Estimation errors

	Mean error	Mean squared error
*KH*1	0.29	0.56
*KH*2	–0.62	0.76
BME	–0.10	0.25
BME(1,2)	–0.13	0.28

6.7 Synopsis

Popular numerical data processing and mapping techniques used by *TGIS* functions are characterized by prevailing *ad hoc* solutions. When faced with the different types of uncertain information arising in real-world applications, many of these techniques often employ tricks specified for the situation of interest and neglect the exploration of a unified theoretical framework that provides epistemically sound and mathematically rigorous models. The advanced *TGIS* functions discussed in this book address successfully some of the limitations of the existing functions, at both the theoretical and the computational levels.

Chapter 7

The *BME* Computer Library

Immunity to boredom gives the computer the edge
A. Lakein

7.1 Computational *BME* Analysis and the *BMElib*

This chapter provides the *TGIS* specialist with a comprehensive presentation of the *BME-based computer library* known by the abbreviated name:

BMElib

The *BMElib* is an essential tool of *computational BME* analysis. It was used to perform the computations involved in many examples presented in this volume. The library includes a large collection of numerical routines. Many of them refer specifically to the computational *BME* approach, although routines dealing with classical space/time statistical analysis and other standard geostatistical operations (variography, spatiotemporal interpolation, and simulation techniques, etc.) are also included in *BMElib*[1].

The reader should not view *BMElib* merely as a cookbook, since its adequate implementation in practice requires a deeper understanding of the underlying *BME* concepts and familiarity with its methods at the level presented in Christakos (2000). It is our hope that the *BMElib* will provide the reader with the tools necessary for integrating/processing knowledge bases and mapping spatiotemporal fields (physical, epidemiological, etc.) in a wide variety of *TGIS* application contexts.

The *BMElib* has been written using the programming language of the *MatLab* software (MathWorks Inc. 1998). Because of this, the reader needs to have access to *MatLab*[2]. In particular, the *BMElib* should be used with *MatLab*-Version 5 or higher. No expertise with any of the various programming languages is required. By presenting *BMElib* as a set of programs within the user-friendly environment of *MatLab*, we expect that the readers will be able to quickly master these programs and even modify them without much effort to fit their own study goals, if necessary.

The present *BMElib* is by no means the final product of computational *BME* analysis. Instead, we view it only as the beginning, since there is a plethora of theoretical *BME* results that have not yet been expressed in a computational form and, as a result, they are not included in the *BMElib*.

[1] Certainly, the *BMElib* has not been developed as a generic geostatistical package, so only the most commonly-used classical techniques are covered.

[2] Providing self-contained source codes in *Fortran*, *C*, or C^{++} might had been more convenient for some users, but these programming languages are typically harder to read, and they lack interactive and graphical capabilities which reduces their usefulness for practitioners who need to adapt the programs to the needs of their specific mapping situations.

7.2 Getting Started

7.2.1 Notational Conventions

Throughout the *BMElib*, all references are entered on the keyboard or printed on the computer screen using a few specific font types. In particular, references to programs, functions etc. are made in bolfaced characters, whereas typing instructions to the reader are given in non-boldfaced characters. Reference to a specific hard data-based *BME* program looks as follows: **BMEprobaMoments.m**. The **.m** suffix always refers to an executable text file called m-files in *MatLab*. Instructions to the reader on how to get detailed information about the function fname is given by typing, help fname.

7.2.2 Getting Started with *MatLab*

The *MatLab* provides a flexible and powerful software especially designed for numerical calculus and matrix computations. Compatible with most major computer systems (Windows, Linux, and Unix), it is widely used in scientific and engineering fields. Although easy to master, *MatLab* offers many powerful features that go beyond strictly numerical applications (e.g., it includes very good graphing capabilities). For general installation instructions and a detailed description of the *MatLab* features, the reader is referred to:

http://www.mathworks.com/.

After starting *MatLab*, the prompt symbol >> appears on the screen. It means that the *MatLab* is waiting for the *BMElib* user to enter a command.

Example 7.1. Type help at the prompt. *MatLab* will provide you with a list of the available functions. To get specific information about a function, type: help fname (fname is the name of the function of interest). ▪

Generally, they are two ways of using *MatLab*. The first one is as a desktop calculator: the user enters a command or a set of commands at the prompt line, executes them, and gets the results. This operation is repeated for each new command. When numerous repetitive commands have to be executed, the job gets cumbersome. The second way circumvents this problem by creating an executable text file (m-file) containing these commands, which can then be readily executed by typing the name of the script file. Both ways can be used to call the numerical *BMElib* routines. These routines are all m-files (or, more specifically, functions executed by *MatLab*). They are invoked by typing the name of the function followed by one or more variables separated by parentheses. Some examples of script files that have been created for an easier and more efficient processing can be found in the exlib or the tutorlib directories described in the *BMElib* User's Guide of Sect. 7.3–7.12 below.

7.2.3
Getting Started with *BMELib*

In this section, the reader will find starting instructions for the *BMElib*. The first step is to download the library. The second step is to run a program in order to compile *Fortran77* routines (only for Linux/Unix platforms). The final step is to setup the *MatLab* search path.

7.2.3.1
Downloading and Installing BMElib

After the *MatLab* has been installed in the reader's computer, the *BMElib* is downloaded on his computer using the *CD* included in this volume[3]. Instructions to install *BMElib* are given in the *CD* in the file *installingBMELIB.html*. The *CD* provides the source code for the subroutine of the *BMElib* package. Additionally some executable files are necessary in the *mvnlib* directory. Two possibilities exist to obtain these executable files:

- For some operating systems (Windows, Unix-solaris, etc.) the executable files already exist in the *CD* or on the *BMElib* website. In this case the reader simply copy these executable files in the *mvnlib* directory.
- If executable files are not available, they must be created by compiling the source code using a *Fortran* compiler properly configured to work with the *MatLab mex* utility. This is done using the *mvnlibcompile.m* command in the *mvnlib* directory, as explained in the installation instructions on the *CD*.

7.2.3.2
Setting up the Path for BMElib

After installing the software, the reader should be able to start using *BMElib*. The final step is setting up the correct path for library use. Since *BMElib* is organized into several directories that contains programs that interact with each other, all these directories must be included into the *MatLab* search path. To get an idea about the current search path, at the prompt type: path. The result looks like this:

```
/usr/local/matlab5.3/toolbox/general
/usr/local/matlab5.3/toolbox/matlab/ops
/usr/local/matlab5.3/toolbox/matlab/lang
/usr/local/matlab5.3/toolbox/matlab/elmat
...
```

These are the directories scanned by default when *MatLab* is searching for a function or a script file. In order to add the new *BMElib* directories into the search path, use the addpath command of the *MatLab*.

[3] The Website http://www.unc.edu/depts/case/BMELIB will also contain relevent information about the installation procedure of *BMElib*. It is recommended to consult this website in case of problems with the installation procedure.

- If the reader is using Windows 98 and *BMElib* is installed in the directory C:\package\BMELIB\, type the following at the *MatLab* prompt:
  >> addpath C:\package\BMELIB\iolib -end
  >> addpath C:\package\BMELIB\graphlib -end
  >> addpath C:\package\BMELIB\modelslib -end
  >> addpath C:\package\BMELIB\statlib -end
  >> addpath C:\package\BMELIB\bmeprobalib -end
  >> addpath C:\package\BMELIB\bmeintlib -end
  >> addpath C:\package\BMELIB\bmehrlib -end
  >> addpath C:\package\BMELIB\simulib -end
  >> addpath C:\package\BMELIB\genlib -end
  >> addpath C:\package\BMELIB\mvnlib -end
  >> addpath C:\package\BMELIB\tutorlib -end
  >> addpath C:\package\BMELIB\exlib -end
  >> addpath C:\package\BMELIB\testslib -end0
- If the reader is using Linux or Unix and *BMElib* is installed in the directory /home/myhomedirectory/BMELIB/, use the same commands as above, but substitute all C:\package\BMELIB\ with /home/myhomedirectory/BMELIB/.

Note that the exact syntax for the command addpath depends both, on the directory where the *BMElib* library is installed, as well as on the platform one is using (see documentation relevant to your system by typing: help addpath). As this operation must be done each time the user is opening a new *MatLab* session, it is convenient to automate the task by putting the corresponding commands into a startup.m text file executed each time the user opens a *MatLab* session.

An example of startup.m text has been created in the *CD*. The reader will have to edit that file and set the correct directory names accordingly. When this is done, he can open a new *MatLab* session, change directory to where the startup.m file is located, and type the command startup at the *MatLab* prompt. The following message should appear on screen:

BMELIB search path has been set.

The reader can then check if the path is correctly setup by typing: help iolib. A list with brief explanations of the functions available in the iolib directory of the *BMElib* should display on screen. For detailed instructions about the use of the file startup.m, see the relevant documentation in *MatLab*. It is recommended that the reader does not work with *MatLab* inside the *BMElib* directory, since there is always the risk of overwriting or deleting essential files. Indeed, the reader can work from any of his own directories and call up *BMElib* functions after the search path for the library has been setup.

7.2.3.3
Structure of the BMElib

The *BMElib* contains several directories, which refer to specific sets of routines. The precise content of each directory can be obtained by typing: help dname (at the *MatLab* prompt; dname is the name of the corresponding directory). More detailed explanations about the aim of the function and the variables needed for invoking it are obtained by typing: help fname (fname is the name of the function). Below we give a list of the various directories found in different sections.

Exploratory Data Analysis

iolib: Offers functions to read/write data from files in different formats, including hard data, soft probabilistic data, *BME* parameters, and some *TGIS* line formats.

graphlib: Once data is entered (from a file or manually), graphlib offers functions to produce graphics that can perform an exploratory analysis of the space/time data patterns.

Space/Time Variability Asssessment

modelslib: Contains several of the most popular covariance/variogram models (exponential, gaussian, nugget effect, spherical, etc.). By combining these models, the reader can create a new space/time model for the mapping situation of his own interest.

statlib: Contains functions for calculating raw covariances/variograms from experimental data, and several other useful statistics (histograms, one-point moments, empirical distributions, etc.).

Spatiotemporal Estimation

bmeprobalib: Useful when hard data are combined with probabilistic soft data. The probabilistic soft data is more general than interval data and can lead to more accurate space/time estimates. The bmeprobalib functions account for multivariate mapping, non-homogeneous/non-stationary trends, and non-Gaussian distributions. In addition, they are flexible in selecting the type of estimator to be used.

bmeintlib: Useful when hard and soft data of the interval type are available. The bmeintlib functions are computationally efficient and account for the same mapping situations as described in bmeprobalib, above.

bmehrlib: Contains space/time mapping algorithms for use when only hard data are available and the general knowledge involves up to 2nd order moments[4]. The bmehrlib functions account for the same mapping situations as bmeprobalib (see above).

S/TRF Simulation

simulib: The simulib functions provide several variants for the two most widely used space/time simulation techniques: Cholesky decomposition and sequential simulation[5]. Both techniques can incorporate hard and soft data.

---

[4] The reader may recall that, when only hard data are available and, assuming certain limiting conditions on the general knowledge, the space/time *BME* estimation algorithm essentially reduces to the *Wiener-Kolmogorov interpolation* or *Kriging* algorithms.

[5] See Christakos (1992).

Miscellaneous Low-Levels Functions

genlib: Contains functions that are of general utility or are invoked by functions in the directories listed above.

mvnlib: Contains all *Fortran77* routines required by the bmeintlib and bmeprobalib mapping functions. These routines perform numerical calculations of multivariate integrals with high numerical cost. Hence, they need to be compiled in order to be optimally efficient. Access to a *Fortran* compiler configured to work with the *MatLab MEX* compiler utility is required.

BMElib Tutorials, Examples, and Tests

tutorlib: Contains tutorials (self explanatory programs) on how to use the functions of the *MatLab* directories.

exlib: Examples in the volume preceded by the mouse icon (🖱) can be reproduced using the programs of this directory (including all numerical results and graphics). These programs are numbered according to the example number in the volume.

testslib: Tests whether a directory is working properly or not. Each test program should end with the statement: test complete (indicating that the test has been successful)[6].

7.2.3.4
Bugs Reporting and Troubleshooting

Typical mistakes include invoking the function with invalid variable definitions (failing to respect the appropriate definitions for the input and output variables will most probably result in a fatal error, when executing the functions). We would greatly appreciate if the readers are kind enough to report to us any major installation troubleshooting or bugs that occurred during the execution of the *BMElib* programs in association with the numerical examples of the book. Although the purpose of the specific *BMElib* version included in the book is to enable the readers to reproduce these numerical examples, the readers are encouraged to contact us about any questions or suggestions they may have regarding the use of the *BMElib*, in general. Also, an effort will be made so that future updates are reported to the public online at our Internet site.

7.2.3.5
BMELIB User's Guide

The Sect. 7.3–7.12 provide a *User's Guide* for the *BMElib*. They are organized according to the main stages of a typical *TGIS* project, as follows:

a Exploratory data analysis.
b Covariance or variogram modelling.
c Spatiotemporal estimation.
d Spatiotemporal simulation.

[6] These programs are useful to advanced users who want to modify some of the *BMElib* functions.

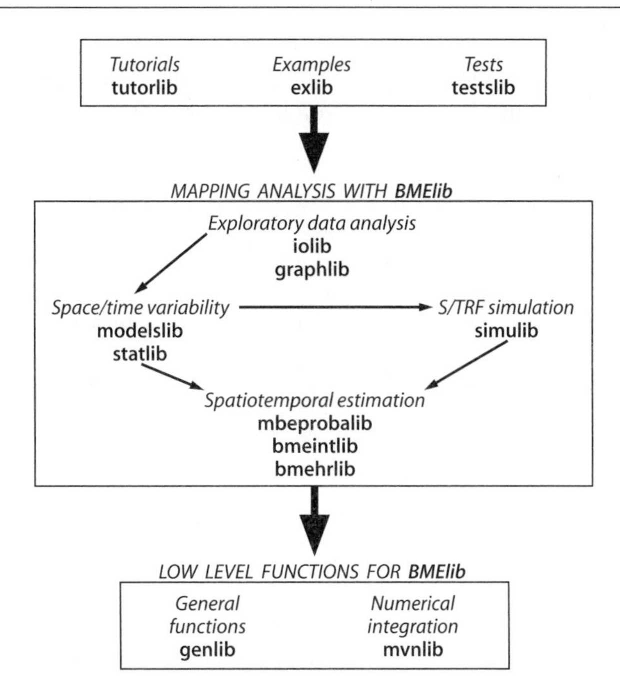

Fig. 7.1. An overview of the *BMElib*

The four main sections above are complemented by additional sections with miscellaneous *BMElib* functions, as well as a section on tutorials, examples and test functions. The User's Guide describes in detail only the most important functions of the *BMElib*. Providing a complete description of all *BMElib* functions would have been unnecessary and cumbersome, because there are currently more than 100 implemented functions. However, online help may be obtained for any of the *BMElib* functions by typing: help fname (fname is the name of the function). It is also possible to get a complete list and brief descriptions of the functions in the *BMElib* directories by typing: help libname (libname is the name of a *BMElib* directory). For most of the functions, the online documentation is sufficient for easy use of the library. An overview of the various directories of the *BMElib* is given in Fig. 7.1. Detailed explanations about the most complex and useful functions in these directories are provided in the following sections.

7.3 The iolib Directory

Exploratory data analysis is an important initial stage of many *TGIS* projects that aim to offer a good grasp of the spatiotemporal behavior of the field (natural or epidemiological) of interest. The *BMElib* functions for exploratory data analysis are organized in two directories:

- The iolib directory for importing/exporting data from files.
- The graphlib directory for visualizing the spatiotemporal distribution of the data values.

This section describes the iolib directory which contains functions that read/write data from files in different formats, including hard data format, soft probabilistic data format, *BME* parameters format, and some *TGIS* line formats.

7.3.1 The readGeoEAS.m and writeGeoEAS.m Functions

These functions can read/write hard data from a file in Geo EAS format. This format has a title in the first line, the number of columns *nc* in the second line, followed by *nc* lines with the name of each column, followed by the data lines (each data line having *nc* values).

Example 7.2. An illustration of a hard data file in Geo EAS format is shown below:

```
BME hard data
3
s1
s2
primary variable Z
10.1     11.4     1.4
12.2     13.6     1.6
16.7     19.1     1.1
10.9     16.9     0.9
```

■

The syntax invoking the **readGeoEAS.m** function to read hard data from a file is:

```
>> [val,valname,filetitle]=readGeoEAS(datafile,colnum);
```

The input variables are as follows:

- **datafile**: A string specifying the name of the datafile from which the user reads the hard data. This file must be in Geo EAS format and have *nc* columns of data values.
- **colnum**: An optional vector specifying which columns to read from the data file. Max(**colnum**) cannot be greater than *nc*. The default value is $1:nc$ (i.e., read columns 1 to *nc*).

The output variables are:

- **val**: An $(n \times nv)$ matrix of values read for the *nv* variables, where $nv = \text{sum}(\text{colnum} > 0)$ is the number of columns read from the data file.
- **valname**: A cell array of *nv* cells with the names of the *nv* variables.
- **filetitle**: A string with the title given in the first line of the data file.

The syntax invoking the **writeGeoEAS.m** function to write hard data to a file is

```
>> writeGeoEAS(val,valname,filetitle,datafile,colnum,outofrangeval);
```

The input variables are:

- **val**: An $(n \times nv)$ matrix of n hard data values for *nv* variables.
- **valname**: A cell array of *nv* cells with the name of the *nv* variables.

- **filetitle**: A string with the title given to the data file.
- **datafile**: A string specifying the name of the generated datafile in which to write the hard data.
- **colnum**: An optional parameter specifying the column assigned to each variable. The default value is 1 : *nv* (i.e., keep the variables in the same order as they are given in the input variable **val**).

7.3.2 The readProba.m and writeProba.m Functions

These functions allow the user to read and write soft probabilistic data from/to a file. Probabilistic data, as defined in Eq. 3.19, refer to the knowledge of the soft pdf $f_S(\boldsymbol{\chi}_{\text{soft}})$ at a set of data points $\boldsymbol{p}_{\text{soft}}$. In *BMElib*, the soft pdf is coded in a discretized form using 4 variables: **softpdftype, nl, limi,** and **probdens.** The reader can type: help probasyntax (for a detailed explanation of how these variables work). The first variable (**softpdftype**) is an integer taking values 1, 2, 3 and 4. It specifies the type of soft pdf as follows (see Fig. 7.2): 1 for histogram, 2 for linear, 3 for histogram on a regular grid, and 4 for linear on a regular grid. An example of the soft pdf $f_S(\chi_{\text{soft}})$ at one datum point is given below.

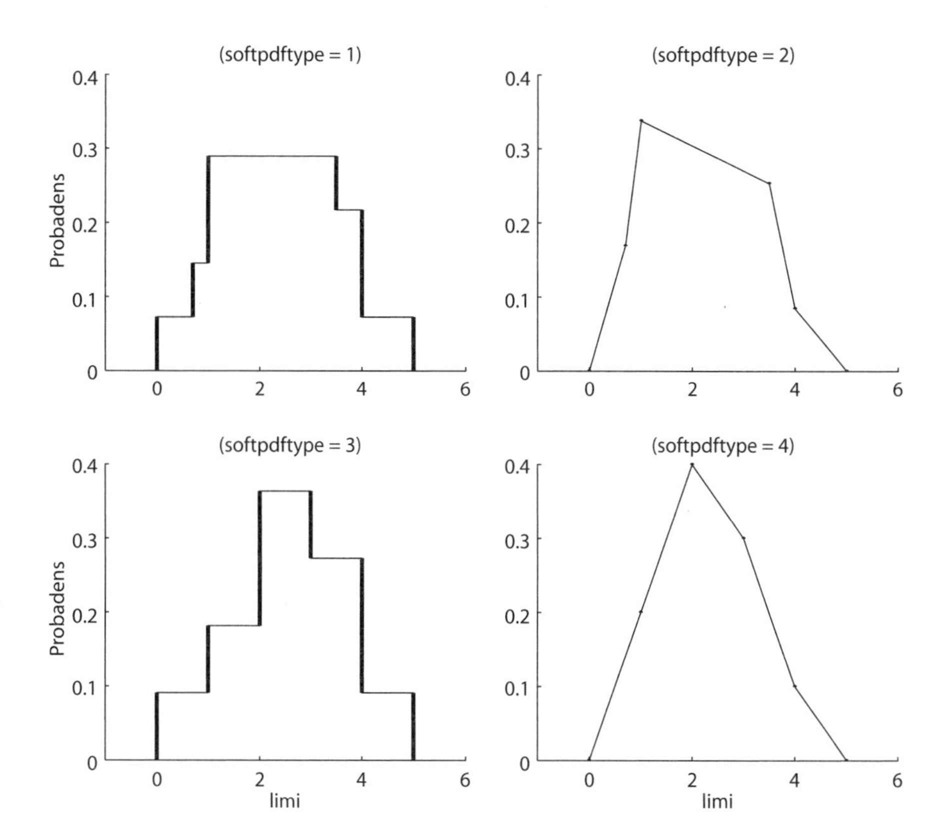

Fig. 7.2. Four types of soft pdf, corresponding to **softpdftype** = 1, 2, 3, 4

Example 7.3. For the subplots on the left side of Fig. 7.2 (**softpdftype** 1 and 3) the soft pdf is discretized as a series of flat steps, whereas it is a series of connected lines for the subplots on the right of Fig. 7.2 (**softpdftype** 2 and 4). The difference between the top row of these subplots (**softpdftype** 1 and 2) and the bottom row (**softpdftype** 3 and 4) is that, on the top row the discretized intervals vary in size, whereas they correspond to a regular grid in the bottom row (i.e., the intervals have a constant size). The second *BMElib* variable **nl** specifies the number of interval limits used to discretize the soft pdf, which corresponds to the number of vertical lines for the histogram type in Fig. 7.2, while it corresponds to the number of dots shown in Fig. 7.2 for the linear types. Finally, the variables **limi** and **probdens** are vectors used to specify discretized $(\chi_{\text{soft}}, f_S)_i$ pair values of the soft pdf $f_S(\chi_{\text{soft}})$. ▪

A file with soft probabilistic data has a title in the first line, the number of column headers *nc* in the second line, followed by *nc* lines with headers, then followed by *ns* data lines. Each data line has the coordinates of the soft data point, a code specifying the variable number, a code specifying the type of the soft pdf (**softpdftype**), the number of interval limits (**nl**), and the vectors **limi** and **probdens** indicating the soft pdf (in discretized form).

Example 7.4. An illustration of a file in such format is shown below:

```
BME Probabilistic data
7
s1
s2
code for the variable (equal to 1)
Type of soft pdf (equal to 1, corresponding to histogram)
number of limit values, nl
limits of intervals (nl values)
probability density (nl-1 values)
  1  0.9   1  1  4    0.1  0.3  0.7  1.1     1.0  1.5  0.5
0.1  0.2   1  1  2    0.1  0.3                5.0
```
▪

The syntax invoking the **readProba.m** function to read some probabilistic data from a file is as follows:

```
>> [cs,isST,softpdftype,nl,limi,probdens,filetitle]=readProba(datafile);
```

The input variable is:

- **datafile**: A string specifying the name of the datafile from which to read the probabilistic data.

The output variables are (see also help for **probasyntax.m**):

- **cs**: An array of two cells, where cs{1} is a $ns \times (d + \text{isST})$ matrix of coordinates, and cs{2} is a $(ns \times 1)$ vector of variable number.
- **isST**: A scalar equal to 1 if the last column of **cs**{1} is time, 0 otherwise.
- **softpdftype**: An integer for the type of soft pdf representing the probabilistic soft data. The integer value specifying the soft pdf type is as follow: 1-Histogram, 2-Linear, 3-Grid histogram, 4-Grid Linear.

- **nl**: ($ns\times 1$) vector of number of interval limits. **nl(i)** is the number of interval limits used to define the soft pdf for soft data point *i*.
- **limi**: ($ns\times l$) matrix of interval limits, where l is equal to either max(**nl**) or 3 depending on the soft pdf type. **limi(i,:)** are the limits of intervals for the i-th soft data.
- **probdens**: ($ns\times p$) matrix of pdf values, where p is equal to either max(**nl**)-1 or max(**nl**), depending on the softpdftype. **probdens(i,:)** are the values of the pdf corresponding to the intervals for the i-th soft data defined in **limi(i,:)**.
- **filetitle**: A string with the title of the file.

The syntax invoking the **writeProba.m** function to write some probabilistic data to a file is as follows:

```
>> writeProba(cs,isST,softpdftype,nl,limi,probdens,filetitle,datafile);
```

The input/output variables are the same as the input/output variables defined above for the **readProba.m** function.

7.3.3 The readBMEproba.m and writeBMEproba.m Functions

These functions are used to read or write from a file the *BME* parameters needed for a space/time estimation. For a description of the syntax of the *BME* parameter files, type: help ioparamsyntax (in *BMElib*); or, see section on the bmeprobalib directory, presented later in this chapter. Using these functions to read and write *BME* parameter files, together with the functions presented earlier to read and write data files, allows the reader to work exclusively from files. To get more information on these functions, type: help readBMEproba and help writeBMEproba.

7.4 The graphlib Directory

Once the hard and soft data have been entered (either from files or manually), *BMElib* offers many graphing facilities that are essential in exploratory data analysis across space and time. The functions in the graphlib directory are designed to visualize data in a 2D space, which is usually the case in *TGIS* applications. For higher spatial dimensions, the reader may still use these functions by selecting a 2D subset of spatial coordinates.

7.4.1 The scatterplot.m Function

This function plots the histograms for a set of fields (natural or epidemiological), as well as the corresponding scatterplots between the fields. The syntax for invoking the function is:

```
>> scatterplot(Z,nbins);
```

The input variables are:

- **Z**: Matrix of values for the different fields (each column is associated with the values of one field). There must be at least two fields.
- **nbins**: Number of bins to be used for drawing the histograms. If **nbins** is omitted from the input list of variables, the function will use the default number of bins as specified in **hist.m**.

7.4.2 The colorplot.m Function

This function plots colored symbols for the values of a vector at a set of 2D coordinates. It uses a colormap so that the color displayed at these coordinates is a function of the corresponding values. The syntax for invoking the function is:

```
>> colorplot(c,z,map,Property,Value);
```

The input variables are:

- **c**: Matrix of point coordinates. Each line corresponds to the vector of 2D point coordinates, so the number of columns is two.
- **z**: Column vector of values to be coded as colors.
- **map**: String that contains the name of the color map to be used. Use help for **graph3d.m** to obtain more information on color maps. E.g., **map**='hot' yields a black-red-yellow-white gradation of colors.
- **Property**: Cell array in which each cell is a string containing a legal name of a plot object property. This variable is optional, as default values are used if **Property** is missing from the input list of variables. Execute: get(H) – H is a plot handle – to see a list of plot object properties and their current values. Execute: set(H) to see a list of plot object properties and legal property values. See also help for **plot.m**.
- **Value**: Cell array where each cell is a legal value for the corresponding plot object property as specified in **Property**.

Example 7.5. Typing, colorplot(**c**,**z**,'hot',**Property**,**Value**), where **Property** ={'Marker', 'MarkerSize','MarkerEdgeColor'}, and, **Value**={'^',10, [0 0 0]}, will plot red triangles with a black border that have a MarkerSize value equal to 10. By default, colorplot(**c**,**z**) will use the 'hot' colormap and circles with a MarkerSize equal to 10 and with a MarkerEdgeColor equal to the default color. ▪

7.4.3 The markerplot.m Function

This function plots the values of a vector at a set of 2D coordinates using symbols of varying sizes, such that the size of the displayed symbols at these coordinates is a function of the corresponding values. The syntax for invoking the function is:

```
>> markerplot(c,z,sizelim,Property,Value);
```

Clearly, most input variables are the same as in **colorplot.m**. An additional variable is:

- **sizelim**: Line vector containing the minimum and maximum values (in pixels) of the size of the symbols to be displayed. The minimum and maximum size values are associated with the minimum and maximum values in **z**, respectively. The size of the symbols for in between values are obtained by linear interpolation.

Example 7.6. Typing, markerplot(**c**,**z**,**sizelim**,**Property**,**Value**), where **sizelim**=[5 20], **Property**={'Marker','MarkerEdgeColor','MarkerFaceColor'}, and, **Value**={'^',[0 0 1],[1 0 0]}, will plot red triangles with a blue border that have a MarkerSize value between 5 and 20 pixels. By default, markerplot(**c**,**z**,**sizelim**), will use disks with the default properties for **plot.m**. ▪

7.4.4 The valplot.m Function

This function plots the values of a vector at a set of 2D coordinates by writing these values as text strings. The syntax for invoking the function is:

```
>> valplot(c,z,Property,Value);
```

Some of the input variables are the same as in **colorplot.m**. In addition:

- **Property**: Cell array in which each cell is a string that contains a legal name of a text object property. This variable is optional, since default values are used if **Property** is missing from the input list of variables. Execute: get(H) – H is a text handle – to see a list of text object properties and their current values. Execute: set(H) to see a list of text object properties and legal property values. See also help for **text.m**.
- **Value**: Cell array in which each cell is a legal value for the corresponding text object property as specified in **Property**.

Example 7.7. Typing, valplot(**c**,**z**,**Property**,**Value**), where **Property**= {'FontSize','Color'} and **Value**={30,[1 0 0]}, will display the text with FontSize=30 and Color=[1 0 0]. By default, valplot(**c**,**z**) will use the default properties for **text.m**. ▪

7.4.5 A Tutorial Use of the graphlib Directory

An example of a set of commands used for visualizing the content of a datafile is given in the **GRAPHLIBtutorial.m** file. For the sake of illustration, some of the commands use optional variables. These variables are not absolutely needed, but they have been incorporated to show how the quality of the output can be easily improved by having control over the colors or the font that is used. The file is executed by typing the following command at the *MatLab* prompt:

```
>> GRAPHLIBtutorial
```

This will automatically start the tutorial. The explanations, the corresponding invoked commands, and the graphics automatically display on the screen.

7.5
The modelslib Directory

The assessment of space/time variability is a key step in a *TGIS* study. The goal is to select a space/time variability model (usually either the covariance or the variogram) for the life support field of interest. *BMElib* includes the most popular models in the modelslib directory. By combining these models the user can construct any space/time model for the mapping situation at hand. These models are fitted to the raw covariance or variogram calculated with the help of the statlib directory (which is described in the next section). A few examples of covariance models are given below.

Example 7.8. Popular spatial covariance models of the modelslib directory are:

1. *Pure nugget effect*, with a variance value c_0.

$$c_x(r) = \begin{cases} c_0, & \text{if } r = 0 \\ 0, & \text{if } r > 0 \end{cases}$$

2. *Exponential*, with a variance c_0 and a range a.

$$c_x(r) = c_0 \exp(-3\frac{r}{a})$$

3. *Gaussian*, with a variance c_0 and a range a.

$$c_x(r) = c_0 \exp(-3\frac{r^2}{a^2})$$

4. *Spherical*, with a variance c_0 and a range a.

$$c_x(r) = \begin{cases} c_0\left(1 - \frac{3}{2}\left(\frac{r}{a}\right) - \frac{1}{2}\left(\frac{r}{a}\right)^3\right), & \text{if } r \leq a \\ 0, & \text{if } r > a \end{cases}$$

■

Several other covariance and variogram models are available in the modelslib directory. The reader can add any custom model, or use any combination of existing standard models in a nested structure.

7.5.1
The *C.m and *V.m Functions

Some of the *BMElib* functions use only covariance models, whereas other functions are able to use both covariance and variogram models. For this reason, covariance and variogram models have been defined in separate functions, namely the *C.m and *V.m functions (the * symbol stands for a string of characters). The spherical model, e.g., is used both as a variogram (**sphericalV.m**) and as a covariance (**sphericalC.m**). Other variogram models do not have a covariance counterpart (e.g., the linear and power variogram models, **linearV.m** and **powerV.m**, respectively). The input variables are as follows:

- **D:** Matrix of distances for which the model has to be evaluated. The distances should be real-valued (≥0).
- **param:** The line vector of model parameters, where the 1st value always corresponds to the covariance sill, or the corresponding parameters for variogram models that do not have a covariance counterpart. In the **linearV.m** function, e.g., the parameter that corresponds to the covariance sill is the slope of the linear variogram model.

The specific parameters for any covariance or variogram model implemented in *BMElib* are obtained by typing: help *C or help *V; the *C or *V is the name of an existing covariance or variogram model, respectively (again, the * symbol stands for a string of characters). The output variable is a covariance matrix **C** or a variogram matrix **V**, depending on the kind of model used.

The reader may easily define additional *customized* covariance or variogram models by duplicating any of the existing functions and modifying the duplicate accordingly. Some conventions have to apply, as follows:

- The 1st input variable of the function must be a distance matrix **D** with real-valued elements (≥0), whereas the 2nd input variable **param** must be a vector of model parameters.
- For the **param** vector, the 1st value must always corresponds to the covariance sill, or its equivalent for variogram models that do not have a covariance counterpart.
- These functions must return a matrix of values that give the elements of the covariance matrix **C** or the variogram matrix **V** at the distances specified in **D** (the **C** and **V** matrices must have the same dimensions as **D**).
- For covariance and variogram models, the last letter of the function name must be C and V, respectively. This allows an automatic syntaxical detection of the case under processing when using some of the *BMElib* functions.
- If the covariance or variogram model is space/time *separable* there is no need to define a new function. For *non-separable* models, the name of the function must end up with the letters ST. If the 1st input variable of the function is a spatial distance matrix **Ds**, the 2nd input variable is a time lag matrix **Dt**, and the 3rd input variable **param** is a vector of non-separable model parameters (**Ds** and **Dt** have the same size). The function must return the elements of the matrices **C** or **V** at the spatial and temporal distances specified by **Ds** and **Dt** (**C** and **V** have the same dimensions as **Ds** and **Dt**). Since distances in space and time are considered separately, it is possible to define customized space/time metrics inside the function itself. For example, the **gaussianCST.m** function of the modelslib directory assumes a non Euclidean space/time metric, where the metric is specified in the **param** vector. Using these tools, several new non-separable models are easily created and added by the user in the modelslib directory.

The reader should make sure that any new covariance or variogram models created are *permissible* in the space/time context considered[7]. The Gaussian model, e.g., is valid only in the case of a Euclidean metric (Example 2.10). When new covariance models are defined by the user, the **nparammodels.m** function must be updated in the proper space/time context. The **nparammodels.m** function provides a list of covariance models in the **models** cell array and the model parameters are listed in the **nparam** cell array. The reader must edit this file and insert the new covariance model names and number of parameters in the corresponding arrays.

When covariance models are used in the *BMElib* functions, the string that contains the name of the covariance model is included in the **covmodel** variable, whereas the **covparam** variable contains the model parameters. When covariance or variogram models are specified, the above variables are named **model** and **param**, respectively. The evaluation of the covariance or variogram matrix between two sets of coordinates **c1** and **c2** is a two step procedure: (*i*) computation of the Euclidean distance matrix **D**, where [D]=coord2dist(**c1**,**c2**); and (*ii*) evaluation of the covariance or variogram matrix **K** associated with **D** using [K]=eval([**model** '(**D**, **param**)']). This procedure is carried out by the **coord2K.m** function using [K]=coord2K(**c1**,**c2**,**model**,**param**). The above covariance and variogram calculations use Euclidian metrics. As long as the reader works with Euclidean distances, he can use the *BMElib* built-in **coord2K.m** function. However, in situations which require the use of a different metric (e.g., the 1-norm metric), the reader can modify the **coord2K.m** to account for the new metric.

7.5.2 The modelplot.m Function

This function can be used to plot the values of covariance/variogram models. This includes nested covariance and variogram models defined as linear combinations of several basic models. The syntax for invoking the function is:

```
>> [v]=modelplot(d,model,param,Property,Value);
```

The input variables are:

- **d:** Column vector of distances (sorted by increasing value) for which the covariance or variogram model must be computed or displayed.
- **model:** String that contains the name of the covariance or variogram model.
- **param:** Line vector of parameters for **model**, according to the conventions assumed for the corresponding covariance/variogram model.
- **Property:** Cell array where each cell is a string containing a legal name of a plot object property. This variable is optional, as default values are used if **Property** is missing from the input list of variables. Execute: get(H) – H is a plot handle – to see a list of plot object properties and their current values. Execute: set(H), to view a list of plot object properties and legal property values (see also help for **plot.m**).
- **Value:** Cell array in which each cell is a legal value for the corresponding plot object property, as specified in **Property**.

[7] For the relevant permissibility criteria, see Christakos (1992) and Christakos and Hristopulos (1998).

Example 7.9. When **Property={'Color','Linewidth'}** and **Value={'[0.5 0.5 1]',1}**, the model is displayed as a purple broken line with a width of 1 pixel. By default, **modelplot.m** will use the default properties for **plot.m**. ▪

If nested models must be displayed, the **model** variable is a cell array such that each cell is a string containing the name of a model. In that case, the **param** variable is a cell array too, where each cell contains the parameters of the corresponding model.

Example 7.10. For a nested variogram model that includes a nugget effect and an exponential model, we have **model={'nuggetV','exponentialV'}**. If the nugget effect is equal to 0.2 and the exponential model has a sill = 0.8 and a range = 1, then we have **param={0.2,[0.8 1]}**. ▪

When an output variable is specified, the graphic is not displayed and **modelplot.m** returns the values for this variable. This optional output variable is:

- v, Column vector of estimated covariance/variogram values at distances specified in **d**.

7.5.3 A Tutorial Use of the modelslib Directory

A tutorial file, **MODELSLIBtutorial.m**, illustrates the use of the functions in the modelslib directory. This file can be executed by typing the name of the file at the *MatLab* prompt.

7.6 The statlib Directory

The statlib directory complements the modelslib directory in the analysis of composite space/time variability. The statlib directory has functions for calculating raw covariances/variograms, as well as several other useful statistical functions for calculating standard distributions (pdf and cdf), empirical histograms, statistical moments, etc. Below, following the list of the functions of this directory, we give the description of only the most important of them as related to the calculation of raw covariances from experimental data.

- Analyzing the distribution of a field (sample histogram, pdf/cdf and other statistics):

decluster.m	Spatial declustering weights.
histline.m	Density-scaled histogram plotted using lines.
histscaled.m	Density-scaled histogram plotted using bars.
cdfest.m	Experimental cumulative distribution.
kerneldensity.m	Pdf estimation using a Gaussian kernel.
cdf2pdf.m	Compute the pdf from the cdf.
skewness.m	Experimental skewness coefficient.
kurtosis.m	Experimental kurtosis coefficient.
spearman.m	Experimental Spearman's correlation matrix.
values2rank.m	Transform a matrix of values into a matrix of rank.

- Transforming to and from the Gaussian distribution:
  gaussplot.m Gaussian probability plot.
  other2gauss.m Transform from an arbitrary pdf to a Gaussian pdf.
  gauss2other.m Transform from a Gaussian pdf to an arbitrary pdf.
- Operations on discretized pdf:
  pdf2cdf.m Compute the cdf from the pdf.
  pdf2CI.m Compute the confidence set from a pdf.
  pdfconvol.m Convolution of a pdf with a Gaussian pdf.
  pdfprod.m Product of two pdf.
  pdfstat.m Statistics of a pdf.
  quantile.m Quantiles computed from the pdf.
  entropy.m Shannon's entropy of a pdf.
- Estimating sample covariance/variogram from experimental data:
  crosscovario.m Single cross-covariance function calculation.
  crossvario.m Single cross-variogram calculation.
  covario.m Multivariate covariance function calculation.
  vario.m Multivariate variogram calculation.
  pairsplot.m Display pairs of points separated by a given distance interval.
  crosscovarioST.m Space/time cross-covariance calculation for vector data.
  crossvarioST.m Space/time cross-variogram calculation for vector data.
  stcov.m Space/time cross-covariance calculation for space/time grid data.
  stmean.m Space/time mean calculation for space/time grid data.
  stmeaninterp.m Interpolation of calculated space/time grid mean estimates.
- Fitting a variogram model to an estimated sample variogram:
  coregfit.m Fitting of the coregionalization model.
  modelfit.m Single variogram/covariance least squares fitting.
  fminsmodelfit.m fmins subroutine for **modelfit.m**.
- Regression:
  regression.m Parameters estimation in a linear regression model.
  kernelregression.m Prediction using a Gaussian kernel regression method.
  designmatrix.m Design matrix in a linear regression model.
- Some classical pdf:
  exponentialpdf.m Exponential pdf.
  exponentialcdf.m Exponential cdf.
  exponentialinv.m Inverse exponential cdf.
  exponentialstat.m Mean, variance, and median of the exponential pdf.
  gausspdf.m Gaussian pdf.
  gausscdf.m Gaussian cdf.
  gaussinv.m Inverse Gaussian cdf.
  gaussstat.m Mean, variance, and median of the Gaussian pdf.
  gaussbipdf.m Bivariate Gaussian pdf.
  gaussbicdf.m Bivariate Gaussian cdf.
  gaussmultpdf.m Multivariate Gaussian pdf.
  gaussmultcdf.m Multivariate Gaussian cdf.
  loggausspdf.m Log-Gaussian pdf.
  loggausscdf.m Log-Gaussian cdf.
  loggaussinv.m Inverse log-Gaussian cdf.

loggaussstat.m	Mean, variance, and median of the log-Gaussian pdf.
triangularpdf.m	Triangular pdf.
triangularcdf.m	Triangular cdf.
triangularinv.m	Inverse triangular cdf.
triangularstat.m	Mean, variance, and median of the triangular pdf.
fmintriangularinv	fmin subroutine for **triangularinv.m.**
uniformcdf.m	Uniform cdf
uniforminv.m	Inverse uniform cdf.
uniformpdf.m	Uniform pdf.
uniformstat.m	Mean, variance, and median of the uniform pdf.

7.6.1 The kerneldensity.m Function

This function is an implementation of the traditional kernel method for estimating a nonparameteric univariate pdf from a set of data values. The Gaussian kernel is used in this function; it is characterized by a smoothing parameter that corresponds to the variance of the Gaussian pdf. The syntax for invoking the function is:

```
>> [pdfzfile]=kerneldensity(z,zfile,v,w);
```

The input variables are:

- **z**: Column vector of field values.
- **zfile**: Column vector of field values for which the pdf must be estimated.
- **v**: Variance of the Gaussian kernel.
- **w**: Optional column vector of weights for the **z** values. These weights must be positive and must total to 1 (see **decluster.m**). If **w** is not specified, all weights are taken as equal.

The output variable is:

- **pdfzfile**: Column vector of estimated pdf values for the kernel smoothed pdf computed at the **zfile** field values.

7.6.2 The pdf2cdf.m Function

This function computes the values of the cumulative distribution function based on a discrete definition of the corresponding pdf. The integration of the pdf is done using a trapezoidal integration formula. The syntax for invoking the function is:

```
>> [cdf]=pdf2cdf(z,pdf);
```

The input variables are:

- **z**: Column vector of field values.
- **pdf**: Column vector of the pdf values at the **z** values.

The output variable is:

- **cdf**: Column vector of the cdf values at the **z** values.

Since the routine uses a trapezoidal integration scheme, it is recommended to have a finely discretized definition of the pdf with values equal to or close to 0 for the lowest **z** values, so that the neglected probability below the lowest value in **z** is null or close to 0. The cdf value at the lowest **z** value is set = 1/2 (value of the cdf at the following **z** value).

7.6.3 The covario.m Function

This function allows the user to estimate the covariances and cross covariances for a set of fields which are known at the same set of points. The syntax for invoking the function is:

```
>> [d,C,o]=covario(c,Z,cl,method,options);
```

The input variables are:

- **c**: Matrix of coordinates at data points. A line corresponds to the vector of point coordinates, so the number of columns is equal to the space dimension (there is no restriction on space dimensionality).
- **Z**: Matrix of field values. Each line is associated with the corresponding line vector of coordinates in the **c** matrix, and each column corresponds to a different field.
- **cl**: Column vector giving the limits of the distance classes used for estimating the covariances and cross-covariances. The distance classes are open on the left and closed on the right. The lower limit for the first class is = 0.
- **method**: String that contains the name of the method used for computing distances between pairs of locations. **method**='kron' uses a Kronecker product, whereas **method**='loop' uses a loop over the locations. Using the Kronecker product is faster for a small number of points, but may suffer from memory size limitations, since it requires the storage of a distance matrix. The loop method may be used regardless of the number of data points considered, and must be used if an Out of Memory error message is generated. Both methods yield exactly the same estimates.
- **options**: Line vector of optional parameters that can be used if default values are not satisfactory (otherwise this vector can simply be omitted from the input list of variables). **options(1)** displays the estimated variograms, if the value is set to 1 (default value is 0). **options(2)** and **options(3)**, with -90=**options(2)** and **(3)**=90, are the minimum and maximum values for the angles to be considered. The angular class is open on the left and closed on the right (angles are counted counterclockwise from the horizontal axis). Two cases must be distinguished: (*i*) if **options(2)**<**options(3)**, the angles α considered are such that **options(2)**<$\alpha\le$**options(3)**; and (*ii*) if **options(2)**>**options(3)**, the α considered are such that $\alpha>$**options(2)** or $\alpha\le$**options(3)**. Angles can only be specified for planar coordinates, i.e., when the number of columns in **c** is equal to 2. **options(4)** is = 0 if the mean is null and = 1 if the mean is constant but not zero (default value = 1).

The output variables are:

- **d**: Column vector of sorted values of the mean distance between pairs of points that belong to the same distance class.
- **C**: Symmetric array of cells that contains the covariance and cross-covariance estimates for the distance classes specified in **d**. Diagonal cells contain the column vector of covariance estimates, whereas off-diagonal cells contain the column vector of cross-covariance estimates. If **Z** is a column vector (only one field), then **C** is simply a column vector of the same size as **d**.
- **o**: Column vector giving the number of pairs of points that belong to the corresponding distance class.

7.6.4 The crosscovario.m Function

This function is concerned with the estimation of a single covariance or cross-covariance. Unlike the **covario.m** function, **crosscovario.m** is able to deal with the case where the values of the two fields are given for two partially or totally different sets of points. When several covariances or cross-covariances have to be computed for the fields available at the same set of points, using **covario.m** is computationally more efficient than the repeated use of **crosscovario.m**. The syntax for invoking the function is:

```
>> [d,c,o]=crosscovario(c1,c2,z1,z2,cl,method,options);
```

Some of the input/output variables are the same as for **covario.m**. Input variables **c1** and **z1** refer to the 1st field, whereas **c2** and **z2** refer to the 2nd field. Also, the output variable:

- **c**: Column vector of raw covariance or cross covariance values (if **z1** and **z2** are identical, the function computes the covariance).

Note that using crosscovario(**c1,c2,z1,z2,cl,method**) yields the same result as using crosscovario(**c2,c1,z2,z1,cl,method**) both for **method**='kron' and **method**='loop'.

7.6.5 The crosscovarioST.m Function

This function performs space/time covariance or cross-covariance estimation. The function can be used when the values of the two fields are given for two sets of partially or totally different space/time points. The syntax for invoking the function is:

```
>> [ds,dt,c,o]=crosscovarioST(c1,c2,z1,z2,cls,clt,options);
```

Some of the input/output variables are the same as in **crosscovario.m**. In addition:

- **c1**: Matrix of space/time coordinates at points where the values of the 1st field are known. A line corresponds to the coordinate vector at a point, so that the number of columns is equal to the dimension of the space plus 1 (last column refers to the temporal coordinate). There is no restriction on space dimensionality.

- **c2**: Matrix of space/time coordinates for points where the values of the 2nd field are known (same conventions as **c1**).
- **cls, clt**: Column vector giving the limits of the spatial and time lag classes, respectively, used in variogram or cross-variogram estimation. As for **cl**, the classes are open to the left and closed to the right. The lower limit for the first class is = 0.

The output variables are:

- **ds**: Matrix of the mean spatial distance between pairs of points for each of the space/time distance classes defined from **cls** and **clt**. Each line is associated with the corresponding spatial distance class in **cls**, whereas each column is associated with the corresponding temporal distance class in **clt**.
- **dt**: Matrix of the mean temporal distance between pairs of points for each of the space/time distance classes defined from **cls** and **clt** (same conventions as for **cls**).
- **c**: Matrix of raw space/time covariance or cross-covariance values. If **z1** and **z2** are identical, the function computes the covariance (same dimensions as **ds** and **dt**).
- **o**: Matrix of the number of pairs of points belonging to the corresponding distance classes (same dimensions as **ds** and **dt**).

Finally, note that only the loop method has been implemented in this function.

7.6.6 A Tutorial Use of the statlib Directory

A tutorial file, **STATLIBtutorial.m**, illustrates the use of the functions in the statlib directory. This file can be executed by typing the name of the file at the *MatLab* prompt.

7.7 The bmeprobalib Directory

The bmeprobalib directory contains a set of functions that perform estimation using both hard data and soft data of the *probabilistic* type using the *BME* formalism. Probabilistic data are more general and informative than the interval data used in the bmeintlib directory or the hard data used in the bmehrlib directory. Hence, the bmeprobalib directory may lead to more accurate spatiotemporal maps and, thus, it contains some of the most advanced *TGIS* functions available in *BMElib*. The physical *KB* and mapping technique used by the bmeprobalib functions include the following three components:

- $\mathcal{G}$-*KB* = mean and covariance functions.
- $\mathcal{S}$-*KB* = hard data and soft probabilistic data.
- *Mapping* = Vector/Single-point estimation.

Each of the functions in the bmeprobalib directory is intended to be as general as possible, covering various practical situations (trend non-homogeneity/non-stationarity, multivariate situations, nested models, composite space/time prediction, etc.).

These situations are arranged into several routines, depending on the kind of mapping sought. The use of the bmeprobalib functions requires the estimation of multiple integrals handled by *Fortran77* subroutines in the mvnlib directory.

7.7.1 The proba*.m Functions

These functions permit the manipulation of probabilistic data. As defined in Eq. 3.19, the data refer to knowledge of the soft pdf f_S coded in a discretized form using 4 variables: **softpdftype, nl, limi,** and **probdens** (see Sect. 7.3.1 above and Fig. 7.2 for a description of these variables, or use help probasyntax in *BMElib* to obtain a detailed explanation of how these variables work). What follows is a brief listing of the **proba*.m** functions in the bmeprobalib directory, permitting the manipulation of the 4 variables above:

proba2interval.m	Transform soft probabilistic data to soft interval data.
proba2probdens.m	Normalize the probability density function.
proba2stat.m	Transform soft probabilistic data to mean/variance data.
proba2val.m	Calculate the value of probabilistic soft pdf.
probacat.m	Concatenate two sets of soft probabilistic data.
probaneighbours.m	Radial neighborhood selection.
probaoffset.m	Offset the soft probabilistic data.
probaother2gauss.m	Transform soft data for arbitrary field, Y, to soft data for Gaussian field, Z.
probasplit.m	Split soft probabilistic data in two sets.
probasyntax.m	Syntaxical help for soft probabilistic data.

7.7.2 The BMEprobaMoments.m Function

The **BMEprobaMoments.m** function computes the statistical moments (mean, variance, coefficient of skewness) of the *BME* posterior pdf at a set of estimation points, using both hard and soft probabilistic data. The mean of the *BME* posterior pdf minimizes the mean square estimation error and is a widely used estimator in mapping applications. The syntax for invoking this function is:

```
>> [moments,info]=BMEprobaMoments(ck,ch,cs,zh,softpdftype,nl,limi,...
   probdens,covmodel,covparam,nhmax,nsmax,dmax,order,options);
```

The input variables are:

- **ck**: Matrix of the estimation points coordinates. A line corresponds to the vector of coordinates at that point so that the number of columns corresponds to the space dimensionality (there is no restriction on the dimensionality).
- **ch**: Matrix of coordinates of the hard data points (same convention as for **ck**).
- **cs**: Matrix of coordinates of the soft data points (same convention as for **ck**).
- **zh**: Column vector of the hard data at the coordinates specified in **ch**.

- **softpdftype:** Integer value specifying the soft pdf type, as follows: 1-Histogram, 2-Linear, 3-Grid histogram, 4-Grid Linear. (see also help for **probasyntax.m**).
- **nl:** ($ns \times 1$) vector providing the number of interval limits. nl(i) is the number of limits used to define the soft pdf at data point i[8].
- **limi:** ($ns \times l$) matrix of interval limits, where l is equal to either max(**nl**) or 3, depending on the soft pdf type. **limi(i,:)** are the limits of intervals for the ith soft datum[9].
- **probdens:** ($ns \times p$) matrix of pdf values, where p is equal to either max(**nl**)-1 or max(**nl**), depending on the soft pdf type. **probdens(i,:)** are the values of the pdf corresponding to the intervals for the i-th soft datum in **limi(i,:)**[10].
- **covmodel:** String containing the name of the covariance model used in the estimation (see modelslib directory). Variogram models are not available for this function.
- **covparam:** Line vector of parameter values used in **covmodel**, according to convention for the corresponding covariance model.
- **nhmax:** Maximum number of hard data considered in estimation at the points specified in **ck**.
- **nsmax:** Maximum number of soft data considered in estimation at the points specified in **ck**. Since the computation time may increase exponentially with **nsmax**, one may choose not to use more than a few soft data points. It is advisable that the **nsmax** used is less than 20.
- **dmax:** Maximum distance between the estimation point and the hard/soft data points. All hard/soft data points separated by a distance < **dmax** from the estimation point are included in estimation, whereas other data points are neglected.
- **order:** Order of the polynomial mean along spatial axes at the estimation points. For the zero-mean case, NaN is used.
- **options:** Line vector of optional parameters used if the default values are not satisfactory[11]. The values for **options(1)**, **options(2)**, and **options(14)** are those used by the **fmin.m** *MatLab* optimization routine that finds the mode of the pdf (default values are the same as for **fmin.m**). **options(3)** specifies the maximum number of evaluations performed by the *Fortran77* sub-routines during integral calculations (default value is 50 000; this value should be increased if a warning message appears on the screen during computation). **options(4)** specifies the maximum admissible relative error in the calculation of these integrals (default value is 10^{-4}). **options(8)** specifies the number of moments to be calculated (1 to calculate only the mean, 2 for the mean and estimation error variance, and 3 for the mean, estimation error variance and coefficient of skewness). The remaining values for **options(1:14)** are not used.

The output variables are:

- **moments:** ($nk \times 3$) matrix of moments estimates at the nk estimation points. Each row of moments has the *BME* mean, estimation error variance (if **options(8)**>1), and coefficient of skewness (if **options(8)**>2) at one estimation point. A value coded as NaN means that no estimation has been performed at that point due to the lack of data.

[8] See also help for **probasyntax.m**.
[9] See also help for **probasyntax.m**.
[10] See also help for **probasyntax.m**.
[11] Otherwise, this vector can simply be omitted from the input list of variables.

- **info**: Column vector providing information about the computation of estimated values. Possible values are: **info**=NaN, if there is no computation at all (no hard and soft data are available around or at the estimation point); **info**=0, when computation is made using *BME* with at least 1 soft datum; **info**=1, when estimation is dubious due to an integration error above the tolerance specified in **options(3)**; **info**=2, when estimation is dubious because the result did not converge within the maximum number of iterations specified in **options(14)**; **info**=3,when the computation is made using only hard data; **info**=4, when there is one hard datum at the estimation point; and **info**=10, when estimation is dubious due to problems with the integration routine (consult the source code for the relevant *FORTRAN* routine).

It is important that the number of points in each spatial neighborhood determined by **nhmax, nsmax** and **dmax** is sufficient with respect to the selected **order** value. E.g., for **order**=1, there must be at least 1+d points in each neighborhood; for **order**=2, there must be at least 1+2*d points, etc. (d is the space dimensionality). The above are the generic input and output variables for space/time estimation. Specific conventions for some of these inputs depend on the case considered, so that different situations can be easily covered by the same function. These specific conventions are adopted in a standard way whenever it is possible to do so for the space/time estimation functions of *BMElib*, including the functions in the bmeintlib and the bmehrlib directories. Various cases are presented next.

The Univariate Case

The simplest use of the space/time estimation function is when only one field variable is used in the estimation process (*univariate* case). If the covariance model includes no nested structures, the **covmodel** is a simple string containing the name of the covariance or variogram model (e.g., **covmodel**='exponentialC'). In this case, **covparam** corresponds to the vector of parameters for the corresponding model (e.g., **covparam**=[1, 4] for an exponential model with variance = 1 and range = 4). For nested models, one needs to specify the names of the models involved together with their parameters. This can be done by using the *MatLab* cell array notation.

Example 7.11. For a nested model with a nugget effect, the notation becomes, **covmodel**={'nuggetC','exponentialC'}. The **covparam** vector is now, **param**={[0.2],[1, 4]}, where 0.2 is the variance of the nugget effect model. Using this notation, the reader can employ as many nested models as desired. ▪

The Multivariate Case

It is also possible to process several natural and/or epidemiological fields at the same time (*multivariate* case). In this case the reader needs to specify additional tags in the **ck, ch** and **cs** matrices. These tags are provided as vectors of values that refer to the field, the values of which range from 1 to *nv*, where *nv* is the number of fields. If there are 2 fields, e.g., the **indexk, indexh** and **indexs** column vectors must have the same number of rows as **ck, ch** and **cs**, respectively; and they contain values equal to 1 or 2, depending on whether the data point refers to the primary or the secondary field, respectively. These vectors are grouped with the coordinates matrices using the *MatLab* cell array notation, so that **ck**={**ck, indexk**}, **ch**={**ch, indexh**}, and **cs**={**cs, indexs**} are now the correct input variables. If **indexk** is a single scalar value, the same field specified by **indexk** will be estimated at the **ck** coordinates. For the multivariate case, the **nhmax** variable is now a (1×*nv*) vector specifying the maximum number of hard data from each field to be used in estimation. Using the same logic, **order** is now a (*nv*×1) vector specifying the order of the polynomial mean along the spatial axes for each field.

When there is a single covariance or variogram to be specified, **covmodel** is still a string that contains the name of the covariance model. **covparam** is now needed to specify the covariance matrix between the fields. For an exponential model having a range of 1.4, e.g., we have **covparam**={C, [1.4]}, where C is the symmetric nv by nv covariance matrix between the nv fields. The reader may notice that covariance models characterized by a single parameter (e.g., **nuggetC.m**) require that the second cell of the **covparam** variable is the empty [.] matrix. For example, in the case of a nugget effect covariance model, **covparam**={C, [ ]}.

In the case of nested covariance models, the **covmodel** cell array is written in the same way as for similar models in the univariate case, but **covparam** is now a cell array containing as many cells as there are variables. For a nested model including a nugget effect, e.g., the notation becomes **covparam**={{Cn,[ ]}, {Ce,[1.4]}}, where Cn and Ce are the symmetric (nv×nv) covariance matrices associated with the nugget effect and the exponential models (for more explanations see **vario.m**, **covario.m** and **coregfit.m** functions). A nested model in the multivariate case is thus defined as a Linear Model of Coregionalization. On the basis of this notation, the reader can use as many nested models as desired.

The Space/Time Case

For space/time data, the convention is that the last column of the **ck**, **ch** and **cs** coordinate matrices corresponds to the time axis. Since analysis involves a composite space/time model (rather than a purely spatial or a purely temporal model), it is necessary to specify if the data are space/time or not, using specific conventions for the name of the covariance model.

Separable covariance models, which are products of spatial and temporal models, are coded as **covmodel**='covmodelS/covmodelT', where covmodelS refers to the name of the spatial covariance model and covmodelT refers to the name of the temporal covariance model. The function will automatically detect the occurrence of the '/' character inside the string and will process the string accordingly. Any combination of valid covariance models can be used. For the **covparam** vector, the first value in the vector always corresponds to the space/time variance. The remaining values represent the other parameters of the spatial and the temporal covariance models in that order.

Example 7.12. If the spatial covariance model is exponential and the temporal covariance model is spherical, **covmodel**='exponentialC/sphericalC'. **covparam**=[1, 1.4, 0.7] means that the variance of the space/time model is equal to 1, the range of the exponential spatial covariance model is equal to 1.4 and the range of the spherical temporal covariance model is equal to 0.7. ▪

For *non-separable* covariance/variogram models, the name contained in the string must end up with the ST characters. The occurrence of these characters in the string is automatically detected by the function. For the **covparam** vector, the first value must always correspond to the space/time variance. For the multivariate case or in the case of nested covariance models, the conventions are exactly the same as described above for spatial data, whether the space/time model is separable or not.

For space/time data, the **dmax** variable becomes a (1×3) vector, instead of a single scalar value. **dmax(1)** is the maximum spatial distance between an estimation point and existing data points, and **dmax(2)** is the maximum temporal lag between an estimation point and existing data points. Only data locations that respect both conditions are considered in space/time estimation. **dmax(3)** refers to a space/time metric, such that the space/time distance = spatial distance+**dmax(3)** × temporal distance. The definition of this space/time distance is needed for selecting the **nhmax** data points that are closest to the estimation points, i.e., at the shortest space/time distance.

It is possible to specify different orders for the polynomial trend along the spatial axis and the polynomial trend along temporal axis. In the univariate case, the **order** is a (1×2) vector, where the **order(1)** is the order of the polynomial along the spatial axes and the **order(2)** is the order of the polynomial along the temporal axis. In the multivariate case, where there are *nv* different fields, the **order** is a (*nv*×2) matrix, where the 1st and 2nd columns of the **order** contain the order for the spatial and temporal polynomials respectively for each one of the *nv* fields. If the order is entered as a (1×2) vector, the same spatial order corresponding to **order(1)** and temporal order corresponding to **order(2)** will be used for all fields.

The Geometric Anisotropy Case

In some situations, the covariance models may have parameters that change according to the direction of the distance vector. The range, e.g., can be maximum along a specific direction and minimum along the perpendicular direction. If the way the range parameter changes with the angle can be described by an ellipse (in 2D coordinates) or by an ellipsoid (in 3D coordinates), this kind of anisotropy is called *geometric* anisotropy. In order to account for geometric anisotropy, the reader has to apply a coordinate transformation using the **aniso2iso.m** function, so that the **ck** and **ch** coordinates are mapped onto an isotropic space (see Fig. 7.3). The **ck**, **ch** and **cs** vectors

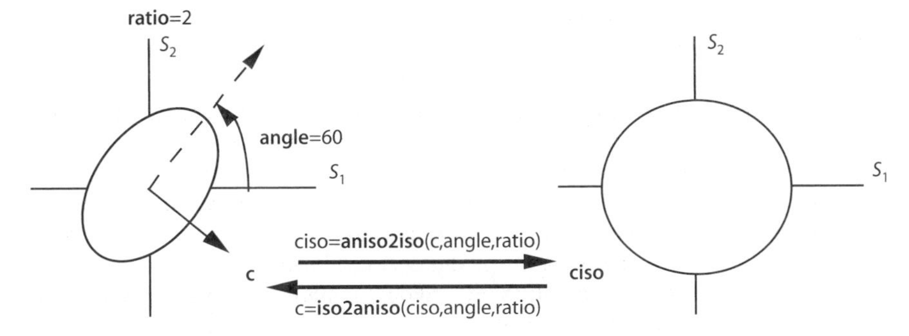

Fig. 7.3. Converting from anisotropic coordinates to isotropic coordinates

are then substituted by the vectors aniso2iso(**ck**, **angle**, **ratio**), aniso2iso(**ch**, **angle**, **ratio**) and aniso2iso(**cs**, **angle**, **ratio**), respectively (see **aniso2iso.m** function in Sect. 7.11.1 for more information about the definition of **angle** and **ratio**). The value for the range of the covariance/variogram model specified in **covparam** is always the range of the model along the principal axis of the ellipse or ellipsoid. It is noteworthy that for nested covariances (i.e., when **covmodel** and **covparam** are cell arrays), the same anisotropy will hold for all specified models.

7.7.3 The BMEprobaMode.m Function

The **BMEprobaMode.m** function computes the mode of the posterior pdf at a set of estimation points, using both hard and probabilistic data. The syntax for invoking this function is:

```
>> [zk,info]=BMEprobaMode(ck,ch,cs,zh,softpdftype,nl,limi,probdens,...
   covmodel,covparam,nhmax,nsmax,dmax,order,options);
```

The input variables and the output variable **info** are the same as in **BMEprobaMoments.m.** In addition, the specific output variable is:

- **zk:** Column vector of estimated mode values. A value coded as NaN means that no estimation is performed at that point due to the lack of available data.

Note that all the specific conventions for specifying nested models, multivariate, or space/time cases are the same as in **BMEprobaMoments.m.**

7.7.4 The BMEprobaPdf.m Function

This function computes the posterior pdf at an estimation point, using both hard and probabilistic data. This function is closely related to the **BMEprobaMode.m** function, but the pdf is obtained at the estimation point, instead of merely the mode. The syntax for invoking this function is:

```
>> [z,pdf,info]=BMEprobaPdf(z,ck,ch,cs,zh,softpdftype,nl,limi,probdens,...
   covmodel,covparam,nhmax,nsmax,dmax,order,options);
```

Some of the input variables are the same as for **BMEprobaMode.m.** The specific input variables are:

- **z:** Column vector of field values for which the pdf must be computed, and sorted in ascending order. **z** may be empty, i.e. **z**=[ ], in which case a grid of **z** values is automatically constructed (see **options(6)** and **options(7)** below).
- **ck:** Line vector of coordinates at the estimation point. Only one estimation point can be considered at a time.

- **options:** Same line vector of optional parameters as for the **BMEprobaMoments.m** function defined above, with the addition of **options(6)** used when **z=[ ]** to determine how fine a grid of **z** values to construct (default value is 25; increase this value to construct a finer grid), and **options(7)** used when **z=[ ]** to determine the extend of the grid of **z** values to construct (default value is 0.001; decrease this value to widen the range covered by the **z** grids).

The **info** output variable is the same as for the **BMEprobaMode.m** function. In addition:

- **z:** Vector of the values for which the posterior pdf is computed.
- **pdf:** Vector of the same size as **z** with the corresponding values of the *BME* posterior pdf.

Note that all the specific conventions for specifying nested models, multivariate, or space/time cases are the same as in **BMEprobaMoments.m**.

7.7.5 The BMEprobaCI.m Function

This function computes confidence intervals (CI) and the *BME* posterior pdf at a set of estimation points, using both hard data and probabilistic data. In the simple case, the CI is just one interval with lower bound **zlCI** and upper bound **zuCI**. But, in general, the CI may be the union of several disconnected intervals, in which case **zlCI** and **zuCI** store the minimum of the lower bounds and maximum of the upper bounds of these intervals, respectively. The syntax for invoking the function is:

```
>> [zlCI,zuCI,pdfCI,PCI,z,pdf]=BMEprobaCI(ck,ch,cs,zh,softpdftype,nl,...
   limi,probdens,covmodel,covparam,nhmax,nsmax,dmax,order,options);
```

Most of the input variables are the same as for **BMEprobaMode.m**. In addition:

- **options:** Same line vector of optional parameters as for the **BMEprobaPdf.m** function defined above, with the addition of **options(20:29)** to indicate the confidence probabilities for which to compute the CI. The CI are calculated only for **options(20:29)** which are ≥ 0.01 and ≤ 0.99. The default value is **options(20)** = 0.68 and **options(21:29)** = 0.

The output variables are as follows:

- **zlCI:** ($nk \times nCI$) matrix storing the lower bound of the CI, where nk is the number of estimation points, and nCI the number of confidence probabilities specified in **options(20:29)**.
- **zuCI:** ($nk \times nCI$) matrix storing the upper bound of the CI.
- **pdfCI:** ($nk \times nCI$) matrix of pdf values at the bounds of the CI.
- **PCI:** ($1 \times nCI$) vector of the probabilities of the CI.

- **z**: $(nk \times 1)$ cell of vectors storing the grid of z values defining the *BME* posterior pdf at each of the nk estimation points.
- **pdf**: $(nk \times 1)$ cell of vectors storing the *BME* pdf value for each one of the **z** grids.

Note that all the specific conventions for specifying nested models, multivariate, or space/time cases are the same as in **BMEprobaMoments.m.**

7.7.6 The BMEprobaTMode.m, BMEprobaTPdf.m, and BMEprobaTCI.m Functions

These three functions calculate the *BME* mode, pdf, and confidence intervals, respectively, when data cannot reasonably be assumed as Gaussianly distributed. In order to deal with a non-Gaussianly distributed natural field, these functions perform, first, a transformation of the data to make them at least marginally Gaussianly distributed, then, second, perform the calculation in the Gaussian domain and, finally, back-transform the calculated output to the original domain. The syntaxes for invoking these functions are, respectively:

```
>> [yk,info]=BMEprobaTMode(ck,ch,cs,yh,ysoftpdftype,ynl,ylimi,yprobdens,...
   covmodel,covparam,nhmax,nsmax,dmax,order,options,yfile,cdfyfile);
```

```
>> [y,ypdf,info]=BMEprobaTPdf(y,ck,ch,cs,yh,ysoftpdftype,ynl,ylimi,...
   yprobdens,covmodel,covparam,nhmax,nsmax,dmax,order,options,...
   yfile,cdfyfile);
```

```
>> [ylCI,yuCI,pdfCI,PCI,y,pdf]=BMEprobaTCI(ck,ch,cs,yh,ysoftpdftype,ynl,...
   ylimi,yprobdens,covmodel,covparam,nhmax,nsmax,dmax,order,...
   options,yfile,cdfyfile);
```

By looking at the function syntax, one can see that these functions have the same input variables as **BMEprobaMode.m, BMEprobaPdf.m,** and **BMEprobaCI.m,** respectively, with these exceptions: there are two additional input variables at the end (**yfile** and **cdfyfile**), as well as two domains (the input data fields are at the original domain, whereas the covariance model for the data is transformed on the Gaussian domain). The input variables corresponding to these exceptions are:

- **y**: Input variables (for **BMEprobaTPdf.m** only) providing the column vector of field values for which the *BME* posterior pdf is computed in the original (non-Gaussian) domain, sorted in ascending order. **y** may be empty, i.e. **y**=[ ], in which case a grid of **y** values is automatically constructed using a coarseness and span determined by **options(6)** and **options(7).**
- **yh:** Column vector of values on the original (non-Gaussian) domain for the hard data at the coordinates specified in **ch.**
- **ysoftpdftype:** Integer for probabilistic soft data in the original domain.
- **ynl:** $(ns \times 1)$ vector of numbers of interval limits for the soft data in the original (non-Gaussian) domain.

- **ylimi:** ($ns \times l$) matrix of interval limits for soft data in the original domain, where l is equal to either max(**nl**) or 3 depending on the soft pdf type (see also help for **probasyntax.m**).
- **yprobdens:** ($ns \times p$) matrix of pdf values for soft data in the original domain, where p is equal to either max(**nl**)–1 or max(**nl**), depending on the softpdftype.
- **covmodel:** String containing the name of the covariance model for the transformed (Gaussian) field. If this model has to be estimated from the data, the data must be transformed to the Gaussian domain prior to estimation (see **other2gauss.m** in statlib directory).
- **yfile:** Column vector of field values for which the cdf is provided in the original domain, sorted in ascending order. It is recommended to have a finely discretized set of values, so that computations are performed on a fine domain.
- **cdfyfile:** Column vector of values for the cdf at the **yfile** values, descibing the distribution of the field in the original (non-Gaussian) domain.

Since these functions use the derivative of the transformation function, it is recommended that the cdf specified in **cdfyfile** be at least twice differentiable. This property assures that the transformed posterior pdf (original domain) is smooth, so that its mode can be easily identified. It is not advisable to use an empirical cdf as provided by the **cdfest.m** function, which is expected to be not smooth. Instead, an estimate of the cdf as provided by **kernelsmoothing.m** or a discrete definition of a differentiable parametric distribution is preferred.

The output variables are the same as for the corresponding functions without transformation, except that they are for the data in the original (non-Gaussian) domain. For example, the output variables **y** and **ypdf** for **BMEprobaTPdf.m** are values of the field and its *BME* posterior pdf in the original scale. If there are several fields to be processed at the same time (see multivariate case described in **BMEprobaMoments.m**), the **yfile** and **cdfyfile** variables must be defined as ($1 \times nv$) cell arrays, where each cell corresponds to the definition of the pdf for the corresponding field.

In the case that there are no hard data available, **ch** and **yh** can be entered as empty [.] matrices. All conventions for specifying nested models, multivariate or space-time cases are the same as in **BMEprobaMoments.m**.

7.7.7 Working with Files

It is possible to work with files in *BME* estimation using hard and probabilistic data. The iolib directory provides functions to read and write the hard data, the probabilistic data, and the *BME* parameters to files[12]. Here we explain the syntax for the file storing the *BME* parameters. A *BME* parameter file is used to store all the *BME* parameters necessary for estimation. This file contains the name of the *BME* method (e.g., *BMEmode* or *BMEmean* method) used to calculate the space/time estimate, the names of the files containing the hard data, the soft data, the coordinates of the estimation points, and all the additional *BME* parameters necessary for space/time estimation.

[12] The syntax for files with hard and probabilistic data have been previously explained.

Example 7.13. Below we show a sample of a *BME* parameter file together with explanations for each of the items in the file. ▪

```
          Parameters for BMEproba
          ----------------------
START OF PARAMETERS:
BMEprobaMode                       \BME method
BMEhard.dat                        \file for hard data
1                                  \nv, number of variables
1 2 0 0 3                          \column for s1,s2,s3,t and var(s)
BMEproba.dat                       \file for probabilistic soft data
0 1.000000e+030                    \data trimming limits
BMEest.dat                         \file with the estimation points
BME1.out                           \output file for BME results
BME1.dbg                           \output file for BME messages
2                                  \number of covariance models
nuggetC                            \covariance model 1
0.05                               \variance for model 1
                                   \additional parameters for model 1
exponentialC                       \covariance model 2
1                                  \variance for model 2
3                                  \additional parameters for model 2
4                                  \max number of hard data
4                                  \max number of soft data
100                                \max search radius
-1                                 \mean trend order(-1 no mean trend)
1000000 0.02                       \maxpts and rEps for numerical integ
0.0001 100                         \relative tol. and maxeval for fmin
50 0.001                           \options(6) and options(7) for pdf
3                                  \options(8) number of moments
0.25 0.5 0.75 0.9                  \CUp tpo 10 Conf. Interval Prob.
0                                  \dotransform, 1 for transf., 0 ow.
```

The first 3 lines of the *BME* parameter file are used for comments only. Each one of the following lines provides input variables, followed by comments (if any). The comments should be separated from the input variable using a few blank spaces, as shown above. The input variables at each one of the lines is explained next (for more detailed explanations use the command help ioparamsyntax in *BMElib*).

BME method:	*BME* techniques use one of the following: 'BMEprobaMode', 'BMEprobaTMode', or 'BMEprobaMoment'.
hard data file:	Name of the hard data file in GeoEAS format.
nv:	Number of field variables (primary+others).
col. numb:	Column number for the *s*1, *s*2, *s*3, *t* coordinates and the field variables.
proba data file:	Name of the file with the soft probabilistic data.
trim limits:	Lower and upper trimming value for the data.
est coord file:	Name of the file with the coordinates of the estimation points in GeoEAS format.
output file:	Name of the file for the output file for the *BME* estimates.
message file:	Name of the file for any messages generated during *BME* estimation.
ncov:	Number of nested covariance models.
covname1:	Name of the first covariance model.
variance1(s):	($nv \times nv$) matrix of variance(s) (and crossvariances if $nv>1$).

param1:	Additional parameters for first covariance model.
covnamei variancei(s) parami	These three input variables are repeated $n$cov times (i.e., i=2,…,$n$cov) in order to specify each of the nested covariance models.
nhmax:	**nhmax**, maximum number of hard data values to be used in estimation.
nsmax:	**nsmax**, maximum number of soft data points to be used in estimation.
dmax:	**dmax**, maximum distance between hard/soft data points and estimation point.
order:	**order**, order of the polynomial drift for the mean trend.
maxpts,rEps:	**maxpts**, **rEps**, parameters for numerical integration.
options(6:7):	**options(6)** and **option(7)**
options(8):	**options(8)** for BME moment estimation.
PCI:	Up to 10 values of Confidence Interval Probabilities (not used).
dotransform:	0 for no transformation, 1 for transformation. When dotransform=1, then you need to specify the vectors yfile and cdfyfile.

The *BME* parameter and data files can be generated manually or using the command **writeBMEproba.m**. When a *BME* parameter file and its corresponding hard and soft probabilistic data files are available, one can easily calculate the *BME* estimates by simply typing **BMEprobaFromFile('BMEparam.dat')**, where **BMEparam.dat** is the name of the *BME* parameter files. This command will read the *BME* parameters and the hard/soft data from files, calculate the *BME* estimates, and will place the result in the output file, thus allowing the user to work directly from files.

7.7.8
A Tutorial Use of the bmeprobalib Directory

A tutorial file, **BMEPROBALIBtutorial.m**, illustrates the use of the functions in the bmeprobalib directory. This file can be executed by typing the name of the file at the *MatLab* prompt.

7.8
The bmeintlib Directory

The bmeintlib directory contains functions that perform estimation using both hard data and soft information of the interval type, according to the *BME* formalism. For soft information of the interval type, it is assumed that the field value at a point belongs to an interval characterized by its lower and upper bounds. Each one of the functions in the bmeintlib directory is intended to be as general as possible, covering various situations (mean non-homogeneity/non-stationarity, multivariate case, nested models, space/time estimation, etc.). The functions are split into several routines, depending on the kind of estimation sought. The use of the bmeintlib functions requires the calculation of multiple integrals handled by *Fortran77* routines in the mvnlib directory (for a detailed discussion of the numerical algorithms used to calculate these integrals, the reader is referred to Serre et al. 1998; and references therein).

7.8.1
The BMEintervalMode.m Function

At a set of estimation points, the **BMEintervalMode.m** function computes the mode (i.e., the most probable value) of the posterior pdf using both hard data and soft interval data. The syntax for invoking the function is:

```
>> [zk,info]=BMEintervalMode(ck,ch,cs,zh,a,b,covmodel,covparam,...
   nhmax,nsmax,dmax,order,options);
```

Some of the input variables and all the output variables are the same as **BMEprobaMode.m.** In addition, the specific input variables are:

- **a:** Column vector of values for the lower bound of the intervals at the coordinates specified in **cs.**
- **b:** Column vector of values for the upper bound of the intervals at the coordinates specified in **cs.**

Note that all the specific conventions for specifying nested models, multivariate, or space/time cases are the same as in **BMEprobaMoments.m.**

7.8.2
The BMEintervalPdf.m Function

This function computes at each estimation point the complete posterior pdf using both hard data and soft interval data. The function is closely related to the **BMEintervalMode.m** function, but the complete pdf is provided at the estimation point, instead of merely the pdf mode. The syntax for invoking the function is:

```
>> [z,pdf,info]=BMEintervalPdf(z,ck,ch,cs,zh,a,b,covmodel,covparam,...
   nhmax,nsmax,dmax,order,options);
```

Some of the input variables are the same as for **BMEintervalMode.m.** The specific input variables are:

- **z:** Column vector of field values sorted in ascending order for which the pdf must be computed. It is recommended to have a finely discretized set of values, so that the pdf can be computed on a fine scale.
- **ck:** Line vector of the estimation point coordinates. For computation at different estimation points, the function is invoked in a loop so that a new **ck** vector is used.
- **options:** Line vector of optional parameters used if the default values are not satisfactory. The **options(3)** and **options(4)** are the same as in **options** for **BMEintervalMode.m.** The values for **options(1)** and **options(2)** are not used.

The output variables are:

- **z**: Possibly modified column vector of values for which the posterior pdf has been computed. The output **z** vector differs from the input **z** vector when there is an interval datum at the estimation point. In that case, the bounds of the intervals are added to the original **z** vector.
- **pdf**: Column vector of pdf values computed at the **z** values.
- **info**: Column vector for information about the computation of estimated values. The different possible values are the same as for **info** in **BMEiprobaMode.m**, except for **info**=2. For **info**=5, the output **pdf** variable is a uniform distribution with bounds equal to those of the interval specified at the estimation location.

Note that when there are no hard data available, **ch** and **zh** can be entered as empty [.] matrices. All the specific conventions for specifying nested models, multivariate or space/time cases are the same as for **BMEprobaMoments.m**.

7.8.3 The BMEintervalTMode.m Function

The **BMEintervalTMode.m** function provides *BME* estimates when the space/time data cannot be assumed to be Gaussianly distributed. Since analysis based on the raw data will not provide very satisfactory results, a solution is first to transform the data so that they become at least marginally Gaussianly distributed, then to perform estimation in the Gaussian domain and, finally, to back transform the estimates in the original domain. The estimates correspond to the mode of the posterior pdf in the original scale. This function is closely related to **BMEintervalMode.m**. The syntax for invoking the function is as follows:

```
>> [yk,info]=BMEintervalTMode(ck,ch,cs,yh,a,b,covmodel,covparam,...
   nhmax,nsmax,dmax,yfile,cdfyfile,options);
```

Some of the input/output variables are the same as for **BMEintervalMode.m**. The specific input variables are:

- **yh**: Column vector of values on the original (non-Gaussian) domain for hard data at the coordinates specified in **ch**.
- **a**: Column vector of values at the original (non-Gaussian) domain for lower bounds of intervals at the coordinates specified in **cs**.
- **b**: Column vector of values at the original (non-Gaussian) domain for upper bounds of intervals at the coordinates specified in **cs**.
- **covmodel**: String containing the name of the covariance model used in estimation (Gaussian scale). If the model is estimated from the data, the data must be transferred to the Gaussian domain prior to estimation (see **other2gauss.m** in statlib directory).
- **yfile**: Column vector of values sorted in ascending order for which the cdf is available at the original scale. A finely discretized set of values is recommended so that computations are performed in a fine scale.
- **cdfyfile**: Column vector of cdf values at the **yfile** values.

The requirement that the **cdfyfile** be a twice differentiable cdf is the same as for the **BMEprobaT*.m** functions, and the same specifications apply.

The additional output variable is:

- **yk**: Vector of estimated mode values at estimation points (original scale). A value coded as NaN means that no estimation was made at this point, due to lack of data.

To use **BMEintervalTMode.m** the field must have a constant mean. Otherwise, one must remove first the values of the mean from the **yh** vector. This is done, e.g., by the **regression.m** or **kernelregression.m** functions of the statlib directory. If several fields are processed at the same time (see multivariate case in **BMEprobaMoments.m**), the **yfile** and **cdfyfile** variables must be defined as ($1 \times nv$) cell arrays, where each cell is associated with the pdf definition of the corresponding field. Again, when no hard data are available, the **ch** and **yh** are entered as empty [.] matrices. All conventions for specifying nested models, multivariate, or space/time cases are the same as for BMEprobaMoments.m.

7.8.4
The BMEintervalTPdf.m Function

This function computes the complete posterior pdf at each estimation point, using hard and soft interval data. The function includes a transformation of the variables. It is closely related to the **BMEintervalTMode.m** function, but a complete pdf is provided at the estimation point instead of merely the mode. The syntax for invoking the function is:

```
>> [y,pdf,info]=BMEintervalTPdf(y,ck,ch,cs,yh,a,b,covmodel,covparam,...
   nhmax,nsmax,dmax,yfile,cdfyfile,options);
```

Some of the input variables are the same as for **BMEintervalTMode.m**. In addition:

- **y**: Column vector of field values in the original (non-Gaussian) domain for which the pdf must be computed and sorted in ascending order. It is recommended to use a finely discretized set of values, so that the pdf can be computed on a fine scale.
- **ck**: Line vector of coordinates at the estimation point. When estimating at different points, the function can be invoked in a loop so that a new **ck** vector can be used.

The output variables are:

- **y**: Possibly modified vector of field values for which the updated pdf was computed. Output **y** vector differs from input **y** vector if there is an interval datum at the estimation point. In that case, the bounds of the intervals are added to the original **y** vector.
- **pdf**: Vector of pdf values computed for the **y** values. The **pdf** values associated with **y** values that are outside the definition of **yfile** and **cdfyfile** are coded as NaN's.
- **info**: Column vector for information about the computation of estimated values, as defined for **BMEintervalPdf.m**.

All remarks made for **BMEintervalTMode.m** are valid, and all conventions for specifying nested models, multivariate, or space/time cases are the same as for **BMEprobaMoments.m**.

7.8.5
A Tutorial Use of the bmeintlib Directory

A tutorial file, **BMEINTLIBtutorial.m**, illustrates the use of the functions in bmeintlib directory. This file can be executed by typing the name of the file at the *MatLab* prompt.

7.9
The bmehrlib Directory

One of the main goals of *TGIS* is spatiotemporal mapping using combinations of hard and soft data. When only hard data are available (and under a set of limiting conditions on the general knowledge considered), spatiotemporal *BME* mapping reduces to the well-known *Wiener-Kolmogorov* and *Kriging* methods. Hence, in the bmehrlib directory we provide several classical Wiener-Kolmogorov and (linear) Kriging mapping algorithms. The bmehrlib directory should be used when *only* hard data are available and the general knowledge involves merely 1st and 2nd order space/time moments. When soft data of the interval type are available in addition to the hard data, one should use the computationally efficient *BME* functions of the bmeintlib directory. If soft probabilistic data are available, one should use the *BME* functions of the bmeprobalib directory[13].

In addition to the classical Wiener-Kolmogorov techniques, some Kriging variations have been included in the bmehrlib as well, like: (*i*) Kriging with measurement errors (*ME*), (*ii*) Kriging with filtered components (*FC*), (*iii*) Kriging with data transformations (*DT*), (*iv*) some restricted forms of multi-point Kriging (*MP*), etc. With the exception of *ME*-Kriging, all other Kriging algorithms can process only hard data. Thus, they have been included mainly for comparison purposes. In general, the *BME* algorithms are capable of processing both hard data and soft data of various kinds. Only a few functions of the bmehrlib directory are described here; several other functions are found in the online help.

7.9.1
The kriging.m Function

This function offers the standard linear Kriging algorithm for processing hard data that can reasonably be assumed as Gaussianly distributed. The function covers various situations, like non-homogeneity/non-stationarity of the mean, multivariate cases, nested models, space/time estimation, etc. Depending on the case, specific formats are needed for the input variables. The general syntax for invoking the function is:

```
>> [zk,vk]=kriging(ck,ch,zh,model,param,nhmax,dmax,order,options);
```

[13]Probabilistic data are more general and informative than interval data and, hence, they can lead to a more accurate space/time estimation.

All the input variables for this function have been described in **BMEprobaMoments.m.** The output variables are:

- **zk:** Column vector of estimated field values at estimation points. A value coded as NaN means that no estimation has been performed at that point due to lack of data.
- **vk:** Column vector of estimation variances. As for **zk**, a value coded NaN means that no estimation has been performed at the point.

It is important that the number of points in each spatial neighborhood determined by **nhmax** and **dmax** is sufficient with respect to the selected **order** value. E.g., for **order**=1, there must be at least 1+d points in each neighborhood; for **order**=2, there must be at least 1+2*d points, etc. (d is the space dimensionality). Note that all the specific conventions for specifying nested models, multivariate, or space/time cases are the same as for **BMEprobaMoments.m.**

7.9.2 The kriginqfilter.m Function

This function performs basically the same computations as the **kriging.m** function, but the user can specify which model components must be estimated and which ones must be filtered-out from the estimation process (e.g., filtering the components with Kriging is a possibility due to the linearity of the algorithm). Both the mean and the model components can be filtered out. Each component is estimated separately and orthogonally to the others. The sum of separately estimated components corresponds to the estimated values yielded by **kriging.m.** The syntax for invoking the function is:

```
>> [zk,vk]=krigingfilter(ck,ch,zh,model,param,nhmax,dmax,order,...
   filtmean,filtmodel,options);
```

Some of the input variables are the same as for **kriging.m**, above. In addition:

- **filtmean:** Value specifying whether the mean is to be estimated (**filtmean**=1) or to be filtered out (**filtmean**=0).
- **filtmodel:** Value specifying whether the stochastic model component is to be estimated (**filtmodel**=1) or to be filtered out (**filtmodel**=0). If the covariance or variogram model are nested, **filtmodel** is a $(1 \times m)$ vector that has as many elements (with values equal to 1 or 0) as there are elements in the **model** cell array.

The output variables are:

- **zk:** Column vector of the estimated field values for the non-filtered mean and/or model components at the estimation points **ck.** A value coded as NaN means that no estimation has been performed at that point due to lack of data.
- **vk:** Column vector of estimation variances for the estimated mean and/or model components at estimation points (same conventions as for **zk**).

All conventions for specifying nested models, multivariate, and space/time cases, are the same as in **BMEprobaMoments.m.**

7.9.3
A Tutorial Use of the bmehrlib Directory

A tutorial file, **BMEHRLIBtutorial.m**, illustrates the use of the functions in the bmehrlib directory. This file can be executed by typing the name of the file at the *MatLab* prompt.

7.10
Simulations

The *BMElib* simulation functions allow the reader to simulate correlated data across space and time. They cover the case of *non-conditional* as well as *conditional* simulations, using hard and soft data of various kinds. These functions, all located in the simulib directory, offer several variants for the two most widely used simulation methods: Cholesky decomposition, and sequential simulation. Only the two most important functions of this directory are explained next.

7.10.1
The simuchol.m Function

This function implements the traditional non-conditional simulation method based on a Cholesky decomposition of the covariance matrix. This simulation method is especially recommended for generating independently several sets of a limited number of hard values (less than few hundreds). Simulated values are zero mean Gaussianly distributed. The syntax for invoking the function is:

```
>> [Zh,L]=simuchol(ch,covmodel,covparam,ns);
```

The input variables are:

- **ch**: Matrix of coordinates at points where hard data are simulated. A line corresponds to a vector of coordinates at a simulation point so that the number of columns corresponds to the space dimensionality (there is no restriction on dimensionality).
- **covmodel**: String containing the name of the covariance model used in simulation (see modelslib directory). Variogram models are not available for this function.
- **covparam**: Line vector of values for the **covmodel** parameters, according to the convention for the corresponding covariance model.
- **ns**: Number of sets of simulated values required. If the optional **ns** variable is omitted from the input list of variables, a single set of simulated values is generated.

The output variables are:

- **Zh**: Matrix of zero mean Gaussianly distributed simulated hard values at the coordinates specified in **ch**. Each column corresponds to a different simulation, so that

if **ns**=1 or if **ns** is omitted, **Zh** is a column vector of values. Each simulated vector of values is statistically independent from the others.
- **L**: Upper triangular matrix obtained from the Choleski decomposition of the global covariance matrix C for values at the **ch** coordinates, such that C=**L'*L**.

All conventions for specifying nested models, multivariate, or space/time cases are the same as for **BMEprobaMoments.m**.

7.10.2 The simuseq.m Function

This function is the implementation of the sequential non-conditional simulation method based on the iterated use of conditional distributions, so that values are simulated one by one. This simulation method is especially recommended for simulating few sets of numerous hard values (above few hundreds), as it does not require the storage of a global covariance matrix into memory. The **simuseq.m** function is based on the use of the space/time estimation function using hard data. As for **simuchol.m**, the simulated values are zero mean Gaussianly distributed. The syntax for invoking the function is:

```
>> [zh]=simuseq(ch,covmodel,covparam,nhmax,dmax,options);
```

Some of the input variables are the same as for **simuchol.m**. The specific input variables are:

- **nhmax**: Maximum number of previously simulated hard data values considered in the simulation at each new point in the sequence.
- **dmax**: Maximum distance between the current and previously visited simulation points. All points separated by a distance < **dmax** from a simulation point are included in the simulation process (other points are neglected). When previously visited simulation points are not available in the local neighborhood (this is likely to happen at the beginning of the simulation), **dmax** is repeatedly increased until there is at least one value in the neighborhood.
- **options**: Line vector of optional parameters used if default values are not satisfactory (otherwise this vector is simply omitted from the input list of variables). **options(1)**=1 for displaying the simulation point currently processed (default value = 0). **options(2)**=1 for a random selection of the visited simulation locations in the sequence (default value = 1).

The output variable is:

- **zh**: Column vector of zero mean simulated Gaussian distributed hard values at the coordinates specified in **ch**.

All the specific conventions for specifying nested models, multivariate, or space-time cases are the same as for **BMEprobaMoments.m**.

7.10.3
A Tutorial Use of the simulib Directory

Tutorial file named **SIMULIBtutorial.m** illustrates the use of the functions of the simulib directory. This file is executed by typing the name of the file at the *MatLab* prompt.

7.11
The genlib Directory

The genlib directory contains the functions that are of general utility for the reader or are invoked by functions from other directories. Only the most important ones are described here. The reader may refer to the online help for information about other functions of the directory not listed here by typing help fname (fname is the name of the function).

7.11.1
The aniso2iso.m Function

Under certain circumstances, it is necessary to transform a set of 2D or 3D coordinates using axes rotations and dilatations to map an anisotropic spatial domain into an isotropic one (see e.g., Fig. 7.3). Geometric anisotropy is characterized by the angle(s) of the principal axis of the ellipse (in 2D) or ellipsoid (in 3D) and by the ratio(s) of the principal axis length over the lengths of the other axes. Using this function, an ellipse (ellipsoid) is mapped into a circle (sphere) having as radius the length of the principal axis. The transformation consists of a rotation of the axes followed by a dilatation. The syntax for invoking the function is:

```
>> [ciso]=aniso2iso(c,angle,ratio);
```

The input variables are:

- **c**: Matrix of point coordinates in the anisotropic spatial domain. A line corresponds to the vector of point coordinates so that the number of columns corresponds to space dimensionality. Only 2D or 3D space coordinates are processed by this function.
- **angle**: Line vector of the angle values that characterize the anisotropy. In a 2D space, angle is the trigonometric angle between the horizontal axis and the principal axis of the ellipse. In a 3D space, spherical coordinates are used, such that **angle(1)** is the horizontal trigonometric angle and **angle(2)** is the vertical trigonometric angle for the principal axis of the ellipsoid. All angles are measured counterclockwise in degrees and are between -90° and 90°.
- **ratio**: Line vector characterizing the length ratio of the axes for the ellipse (in 2D) or the ellipsoid (in 3D). In a 2D space, **ratio** is the length of the principal axis of the ellipse divided by the length of the secondary axis, so that **ratio**>1. In a 3D space, **ratio(1)** is the ratio of the length of the principal axis of the ellipsoid over the length of the second axis. **ratio(2)** is the length of the principal axis of the ellipsoid divided by the length of the third axis, so that **ratio(1)**>1 and **ratio(2)**>1.

The output variable is:

- **ciso**: Matrix of the same size as **c** that gives the new coordinates in the isotropic space.

It is possible to specify an additional index vector taking integer values from 1 to *nv*. These values specify which one of the *nv* fields is known at each one of the corresponding coordinates. The **c** matrix of coordinates and the index vector are grouped together using the *MatLab* cell array notation, i.e., c={c,index}. This permit a single, simultaneous coordinate transformation on a set of possibly different fields. The output variable **ciso** is also a cell array that contains both the new matrix of coordinates and the index vector.

7.11.2 The iso2aniso.m Function

This function performs a transformation which is reciprocal to that performed by the **aniso2iso.m** function, so that when applied successively both transformations have no effect on the coordinates. Thus, this function maps a set of coordinates of an isotropic space into a set of coordinates in an anisotropic one. The syntax for invoking the function is:

```
>> [c]=iso2aniso(ciso,angle,ratio);
```

The **angle** and **ratio** variables are the same as for **aniso2iso.m**, but now **ciso** is the input matrix of coordinates in the isotropic space and **c** is the output matrix of coordinates in the anisotropic space. The same conventions like those used in **aniso2iso.m** apply here.

7.11.3 The coord2dist.m Function

This function computes the Euclidean distance matrix between two sets of coordinates. The syntax for invoking the function is:

```
>> [D]=coord2dist(c1,c2);
```

The input variables are:

- **c1**: Matrix of coordinates for the points in the first set. A line corresponds to the vector of coordinates at a point, so that the number of columns is equal to the space dimensionality (there is no restriction on space dimensionality).
- **c2**: Matrix of coordinates for the points in the 2nd set, using the same conventions as for **c1**.

The output variable is:

- **D**: Euclidean distance matrix between the coordinates in **c1** and **c2**. The number of lines and columns for D correspond to the numbers of lines in **c1** and in **c2**, respectively.

As for **aniso2iso.m**, it is possible to specify additionally a $(n1 \times 1)$ index1 and a $(n2 \times 1)$ index2 vectors, taking integer values from 1 to nv. The values in the index vectors specify which one of the nv field is known at each one of the corresponding coordinates. The **c1** and **c2** matrices of coordinates and the index vectors are then grouped together using the *MatLab* cell array notation, so that **c1**={**c1**,index1} and **c2**={**c2**,index2}.

7.11.4 The coord2K.m Function

This function computes the covariance or variogram matrix between two sets of coordinates, based on Euclidean distances. The syntax for invoking the function is:

```
>> [K]=coord2K(c1,c2,model,param,filtmodel);
```

Part of the input variables are the same as in the case of **coord2dist.m**. The specific input variables are:

- **model**: String that contains the name of the covariance or variogram model used in computation.
- **param**: Line vector of values for the **model** parameters according to the convention for the corresponding covariance or variogram model (see modelslib directory).
- **filtmodel**: Optional value specifying if the stochastic model component is to be included (**filtmodel**=1) or is to be filtered out (**filtmodel**=0). See **krigingfilter.m** function in the bmehrlib directory.

For a detailed discussion on the coding of the **model** and **param** variables in various situations (e.g., nested models, multivariate case, space/time case), the reader is referred to the **BMEprobaMoments.m** function. The output variable is:

- **K**: Covariance or variogram matrix between the coordinates in **c1** and **c2**, depending on whether **model** represents a covariance or a variogram model. The number of lines and columns for **K** corresponds to the number of lines in **c1** and in **c2**, respectively.

7.11.5 The kernelsmoothing.m Function

This function uses a Gaussian kernel filter to obtain a smooth estimate of a field based on the availability of hard data at a number of points. The function is useful in the mean trend modelling of a non-uniform field. The estimate at a point is a weighted linear combination of hard data at surrounding points. The weights are positive and proportional to the values of a Gaussian distribution evaluated at Euclidean distances between the estimation points and the points where hard data are known. The syntax for invoking the function is:

```
>> [zk]=kernelsmoothing(ck,ch,zh,dsmoothing,nhmax,dmax,options);
```

The input variables are:

- **ck**: Matrix of coordinates at the estimation points. A line corresponds to the vector of coordinates at an estimation point, so that the number of columns in **ck** corresponds to the space dimensionality (there is no restriction on dimensionality).
- **ch**: Matrix of coordinates for the hard data points, with the same convention as for **ck**.
- **zh**: Column vector of values for the hard data at the coordinates specified in **ch**.
- **v**: Variance of the isotropic Gaussian kernel distribution. A higher value for **v** provides a higher smoothing for the **zk** estimates. Three times the square root of **v** is the practical range of the Gaussian kernel smoothing filter, i.e., it is the distance from the estimation point beyond which hard data points have very small associated weight.
- **nhmax**: Maximum number of hard data considered for each estimation point.
- **dmax**: Maximum distance between an estimation point and existing hard data points. All hard data points separated by a distance < **dmax** from an estimation point will be included in the estimation process (other hard data points are neglected). The value of **dmax** should be no greater than 3 times the square root of **v**.
- **options**: Optional parameter that can be used if the default value is not satisfactory (otherwise this parameter can simply be omitted from the input list of variables). **options** is equal to 1 or 0, depending on whether the user wishes or not to display the order number of the point currently processed by the function. Default value = 0.

The output variable is:

- **zk**: Column vector of estimates at estimation points. Values coded as NaN mean that no estimation has been performed at that point, due to lack of data in the neighborhood.

7.11.6 A Tutorial Use of the genlib Directory

Like with all other *BMElib* directories, a tutorial file named **GENLIBtutorial.m** illustrates the use of the functions of the genlib directory. This file is executed by typing the name of the file at the *MatLab* prompt.

7.12 The mvnlib Directory

All *Fortran77* subroutines that are needed for the bmeintlib and bmeprobalib estimation functions are located in this directory. These *Fortran77* sub-routines perform the numerical calculation of multivariate integrals with high numerical cost, thus requiring that they be compiled in order to be optimally efficient. Therefore, the user must have access to a *Fortran* compiler configured to work with the *MatLab MEX* compiler utility in order to compile these functions[14].

[14]The reader is referred to Serre et al. (1998), and references therein for a discussion of numerical efficiency issues and some of the numerical algorithms used.

7.12.1
The mvnlibcompile.m Function

This program is used when installing *BMElib* in order to compile all the *Fortran77* subroutines in the mvnlib directory. For this program to work, the *MatLab MEX* utility has to be correctly configured with an allowable *Fortran* compiler (check the installation guidelines for more explanations). To invoke this program, the user has to first use the cd command to move to the mvnlib directory (this is very important because otherwise the compiled files will be created in the wrong directory), and type the following command at the *MatLab* prompt:

```
>> mvnlibcompile;
```

7.12.2
Testing the mvnlib Directory

In order to test if the mvnlib directory was correctly compiled during the *BMElib* installation, there is a test file named **MVNLIBtest.m** in the mvnlib directory. This file can be executed by typing the name of the file at the *MatLab* prompt.

7.13
BMElib Tutorials, Examples, and Tests

7.13.1
The tutorlib Directory

The tutorlib directory contains *MatLab* script files that are presented as examples of how to use the various *BMElib* directories. These script files are plain text files, so that they can be opened and modified with any text editor. They contain a set of documented *MatLab* commands that automatically call various *BMElib* functions and generate some plots of the corresponding results. These script files are not designed to give an exhaustive overview of the complete sets of functions available in the library. They should rather be considered as guidelines for encoding the data and invoking the major functions that the user is likely to need the most frequently. The name of each file is composed by a prefix which is the name of the directory that the script file is illustrating in capitalized letter, followed by the tutorial.m suffix. The **GRAPLIBtutorial.m** script file, e.g., illustrates the use of the graphlib directory. Any of these files can be executed by typing the name of the corresponding file at the *MatLab* prompt.

7.13.2
The exlib Directory

In addition to the tutorlib directory that contains script files strictly related to the general use of the *BMElib*, the exlib directory contains more objective-oriented script files related to some of the numerical examples that appear in this volume. After installing the *BMElib*, the user can generate again the graphs and the results corresponding to these examples by typing the name of the corresponding script file at the prompt.

These script files are plain text files, and can be edited and modified with any text editor. This should allow the user to get a deeper and systematic insight of the *BME* theory and methodology underlying each the numerical examples.

7.13.3
The testslib Directory

Testing functions are included in this directory that test if each directory is working properly. Each test function executes various tests and ends by printing the statement "test complete" on the screen, thus indicating that the test has been successful. These functions are useful to check whether *BMElib* was correctly installed. In addition, they can be used by advanced users who have modified some *BMElib* functions and want to verify that the *BMElib* is still working properly. The name of each function is composed by a prefix which is the name in capital letters of the directory to be tested, followed by the test.m suffix. The **BMEPROBALIBtest.m** function, e.g., is used to test the bmeprobalib directory. Any of these files can be executed by typing the name of the corresponding file at the *MatLab* prompt.

Chapter 8

Scientific Hypothesis Testing, Explanation, and Decision Making

Understanding is a lot like sex.
It's got a practical purpose,
but that's not why people do it normally.
F. Oppenheimer

8.1 On Scientific Methodology

As it was pointed out on several occasions in the preceding chapters, one of the major epistemic contributions of modern *TGIS* in scientific development is its well-established ability to

Synthesize in a mathematically rigorous and epistemically sound way a variety of KB relevant to the phenomenon of interest, in order to generate highly informative and numerically accurate maps across space and time.

Therefore, *TGIS* can play a pivotal role in the *horizontal integration* of sciences which leads to new highly interdisciplinary fields, many of which lie at the frontier of scientific development. For example, by integrating environmental data with demographics, facility information, health data, and infrastructure characteristics the *TGIS* specialist can address a series of issues concerning the well-being of a region (see discussions in Christakos and Hristopulos 1998; and in Osleeb and Kahn 1999).

Scientific development seeks to make sense of the real world by deriving a set of mental constructions about it. A basic constituent of such an effort is the establishment of an adequate *methodology*, i.e., a coherent step-by-step procedure for thinking critically about scientific development and acting upon it. Hence, in addition to the specific epistemic contribution described above, another *TGIS* characteristic of vital importance in the context of scientific development is that

Spatiotemporal information generated by TGIS *can play a substantial role in scientific methodology.*

In particular, space/time maps could very well serve the goals of scientific methodology in various ways, e.g., to test a hypothesis regarding the origin and dispersion of an epidemic, to explain a cause and effect relationship, to assess the risks to the ecosystem of an exposure to certain environmental pollutants, and to make sound decisions. Before we proceed with a more detailed description of *TGIS*'s role in this context, let us give a brief review of the topic of scientific methodology from a philosophical viewpoint.

Basically, a scientific methodology offers a perspective of reality based on a specific philosophical framework (dictated by ontological and epistemic systems of ideas, assumptions, facts, assertions, beliefs, and theories). As we noticed in the previous chapters,

The philosophical framework within which the TGIS expert choses to operate can have considerable bearing on the various components of the research project

(including its chosen topics, methods of investigation, observation techniques, interpretation of the results, and justification of the conclusions). For example, if the philosophical framework adopted by an expert is realism (i.e., he believes that a real world exists independent of people), he will design the research project in a different way than another expert whose philosophical framework is metaphysical idealism (i.e., he believes that people construct their own world). As should be expected, the very structure of scientific development has been a central topic of discussion and debate among the most eminent philosophers of science. The frameworks proposed by the various schools of philosophy differ because of their different perspectives on what constitutes reality, how knowledge is acquired and represented, what are the appropriate uses of scientific information, etc. What follows is a brief exposition of some of the major philosophical frameworks that are of considerable interest to the *TGIS* researcher (for a more detailed discussion the interested reader is referred to the relevant literature).

Popper (1934) emphasized the logical structure of science and adopted the so-called "falsificationist" framework to study scientific development. According to falsificationism, there is no final truth in science, and the testing of a model or a hypothesis proceeds deductively through attempted refutations (i.e., they are conjectured and then put to the test by examining the observational conclusions derived from them). A hypothesis is accepted – "corroborated" in Popper's terminology – if it has been unrefuted and has stood up to severe testing. Simply put, one recognizes a hypothesis to be valid, when to deny it would be a clear contradiction of demonstrable facts.

Kuhn (1962) introduced a historical perspective in the study of scientific practice and used the term "paradigm" to describe a particular way of looking at things. The paradigm includes a set of theories, laws, techniques, applications, and instrumentation together. The paradigm is closely related to what Kuhn called "normal science", i.e., the dominant framework of actual scientific practice which decides the problems worth studying, the kind of questions to be asked, the theoretical methods to be used and experiments to be performed in attempting to answer these questions, and establishes the peer review process that controls both the boundaries of accepted works as well as their quality[1].

Lakatos (1976) used the term "program" to suggest a methodology of scientific progress that is a combination of certain aspects of Popper's idealism (e.g., the commitment to a rational method) with Kuhn's realism (e.g., considering scientific development from the normal science viewpoint). Lakatos' program emphasizes the consideration of several competing hypotheses, rather than a single one. Among all of them, the one hypothesis is selected that better explains the data available and leads to the discovery of novel aspects of reality.

---

[1] According to Kuhn, normal science is concerned with "puzzle-solving" within the framework of the existing theories. Normal science is not concerned with the testing of global theory, which is the objective of scientific revolutions.

Within the context of scientific methodology, the importance of *prediction* in terms of spatiotemporal maps can hardly be overestimated. These spatiotemporal maps frame observations the *TGIS* specialist would expect to make, assuming that the underlying scientific hypotheses and models are valid. An influential view of scientific development, known as the hypothetico-deductive approach (e.g., Rosenberg 2000), emphasizes the salient role of observable prediction, in conjunction with hypothesis testing and model building:

Scientific development is based on sets of hypotheses, which are tested by logically deriving observable predictions from them. If these predictions are observed, in experiment or other data collection, then the hypotheses which the observations test are tentatively accepted.

This view has not been without opposition in recent years (e.g., Oreskes 2000). Certain social scientists and postmodern philosophers, e.g., have questioned the intellectual worth of predictions[2]. This opposition seems to be based on a misdirected approach towards satisfying the real needs of the world. Whatever the motivation behind the opposition may be, and regardless of ephemeral technical difficulties, nobody can seriously doubt the paramount importance of making predictions. It will be extremely hard, no doubt, to justify the practical utility of many scientific disciplines when they cannot provide predictions. Indeed, in the eyes of the public a science that makes no useful predictions is often little more than armchair philosophy. It is difficult to imagine, e.g., a science of mechanics that cannot predict the paths of cars moving in the highways, airplanes flying from one city to another, and satellites orbiting the earth. Or, a medical science that does not attempt to predict the human body's reaction to specific drugs. An environmental science that has no interest in predicting the effects of certain pollutant exposures on local populations is of little practical value. What is the usefulness of atmospheric and climatic observations, if they cannot lead to predictions about the weather? True, in many cases these predictions are uncertain (as a result of the complicated phenomena involved, the numerous variables influencing the outcome of the analysis, etc.), but it would be inappropriate – to say the least – to dismiss prediction as an important concept of scientific development because of the unavoidable uncertainty of the real-world. Instead, the sensible approach is for the scientists to increase the accuracy of their predictions by continuously improving their understanding of the underlying mechanisms and the uncertainties involved at various levels of the scientific analysis.

Despite their differences regarding certain methodological aspects, most of the above philosophies of science agree on a few crucial issues. As mentioned above, one of these issues is the importance of prediction across space and time (e.g., predicting future outcomes of a natural phenomenon and its effects on local populations). Another one is that, without a methodological framework (paradigm or program) that is acceptable

[2] One group of critics, e.g., doubt the scientific substance of predictions generated by certain computer schemes. These schemes, according to these critics, constitute a complex mixture of phenomenological and empirical laws, input parameters, and auxiliary relationships, without a clear understanding of the conceptual framework at work, or its validity and physical consistency. The interested reader is referred to Sarewitz et al. (2000); and references therein.

within the limits of normal science as currently conceived for the problem under consideration, scientists cannot even gather meaningful "facts"[3]. Furthermore, most serious attempts towards the establishment of a sound scientific methodology that can be applied efficiently in practical situations have come to the same conclusion:

No one philosophy by itself can explain all the complexity of reality.

Indeed, due to the large variety of forms of scientific practice encountered in life support sciences, no unified program or paradigm is yet available to capture such a practice in all its variety (large-scale developments leading to major paradigm shifts vs. local-scale details associated with particular scientific disciplines, etc.). Instead, several partial frameworks exist, each one of which has a different breadth of scope and describes broad or specific facets of scientific practice. A scientists may find that some aspects of his investigation point to one program and others to different programs. Hypothetico-deduction can play a crucial role in judging alternative theories and models; induction-based analogies are important in developing useful hypotheses; comparative forms offer a rigorous basis for choosing between a series of competing scientific hypotheses; etc. In the case of *TGIS*, the determination of an adequate methodology is a continuously involving multi-scale and multi-purpose project, which involves one or more scientific disciplines. Nevertheless, for most practical purposes many *TGIS* specialists find it constructive to employ a scientific methodology that consists of the following three distinct stages:

1. The *classification* stage (in which data are collected and tabulated, the basic terminology of the subject is created, etc.).
2. The *correlation* stage (in which certain important correlations between data are detected, practical interpolations within the data domain based on past experience are made, etc.).
3. The *cause-effect* stage (in which powerful hypotheses are derived that explain existing correlations and discover new associations, logical derivations based on assumed causes predict the outcomes of new situations, etc.).

As we noticed in Chap. 5 even at the classification stage 1 an expert does not usually start with blank ignorance of what the phenomenon in question represents. Instead, he often possess a rather vague or merely practical grasp of the situation, and he seeks to pass to an explicit and reasoned grasp of what the natural phenomenon is (which is the goal of stages 2 and 3). From an epistemic viewpoint, scientists often distinguish between *deterministic* cause-effect and *stochastic* cause-effect situations (for a more detailed discussion, see Sect. 8.3 below). In the view of many *TGIS* specialists the third stage above is by far the most important stage of scientific methodology[4]. Scientific hypothesis and explanation constitute part of the terminology of the cause-effect stage 3

[3] According to Kuhn, e.g., at the earlier stage (sometimes called the preparadigmatic stage) the gathering of facts is made by several groups of researchers who may be not even sure what count as important data and what data are meaningful.

[4] In the cause-effect stage the center of interest are questions of the "Why"-type, whereas the correlation stage is concerned with questions of the "How"-type.

(this terminology will be the topic of the following two sections). At this stage, any unimportant data gathered at the classification stage are filtered out, whereas the explanatory relevant facts are organized among them into a theoretically useful hierarchy. We now turn our attention to the discussion of scientific hypothesis testing.

8.2 Hypothesis Testing

Let us start this section with a general definition of the very important subject of hypothesis testing. In many cases, the goal of the research project is to assess the validity of a hypothesis. For the *TGIS* purposes of this book, a useful descriptive definition of scientific hypothesis is as follows:

A scientific hypothesis is an unproved proposition about a specific phenomenon based on our current understanding of the situation, which is assumed to explain certain aspects of the phenomenon and offer the basis for further investigation.

Furthermore, a hypothesis may involve more than one phenomenon or natural variable (e.g., it could be a conjectured relationship between several phenomena or natural variables). A scientific hypothesis, in the above sense, has certain important features that distinguish it from other kinds of hypotheses, such as a *statistical* hypothesis. The latter typically refers to a distinct characteristic of a narrowly defined population (e.g., a sample survey or a set of repeated measurements of a variable) and uses formal tests of statistical inference which make no reference to the underlying physical mechanisms and laws[5]. As a result, the pure induction-based projection of observed regularities into the future could be meaningless (e.g., there may exist a multitude of statistical regularities present in any sequence of observations) unless one makes sure that these regularities are physically consistent with the phenomenon under consideration. By way of a summary,

The testing of a scientific hypothesis is a tedious process that involves experimentation, physical modelling and epistemic analysis that go far beyond the purely evidential techniques of statistical inference.

This important point has been emphasized by researchers in various scientific disciplines. The fact that statistical correlation does not necessarily imply scientific causation is widely appreciated in sciences (e.g., Armstrong 1983). Also, Hilborn and Mangel (1997) concluded that statistical significance often has little, if any, relation to biological significance. In his study of causal modelling, Asher (1983) argued that one should not allow the testing and revising of models to become an enterprise completely determined by statistical results devoid of theoretical underpinnings. Wang (1993) observed that, the scientific community does not accept the evolution theory and rejects the creation theory on the basis of formal statistical inference (e.g., by simply

[5] As a result, in the eyes of scientists several tests and procedures of classical statistics lack a coherent rationale (e.g., their interrelations are not obvious, and offer little physical insight regarding which test to use and why; Sivia 1996).

setting up some null vs. alternative hypotheses and conducting some kind of Neyman-Pearson hypothesis testing). Scientific hypothesis testing is often a much deeper affair than simply defining some positive instances[6]. In most cases, the validity of a hypothesis cannot be completely established in terms of observational evidence. Also, a single falsifying test cannot determine with certainty whether the problem is with the hypothesis under test or with the *auxiliary* assumptions that have been made in order to uncover the falsifying evidence.

Often, a *TGIS* specialist may find it preferable to view scientific hypothesis testing as a process in which, instead of simply examining statements of the form "An event A contributes to the occurrence of a phenomenon B in space and time", one should pose more inquisitive questions of the type "What are the possible ways in which an event A can contribute to the occurrence of a phenomenon B across space and time?" In such a context, a scientific hypothesis may involve one or more *models*. Modelling has been proven to be an invaluable tool, e.g., in public health studies involving "what-if" tests and performance indicators (Birkin et al. 1996). In some cases, a set of models may refer to the same physical situation (a subsurface contamination study could use a series of models, all representing multiphase flow in a porous formation), whereas others may be associated with different scientific disciplines (e.g., an environmental study may involve three groups of models, the first one representing atmospheric pollution, the second one describing climatic conditions, and the third model assessing ecological damage).

Example 8.1. The hypothesis that an environmental contaminant (e.g., arsenic) is the cause of a specific disease (e.g., some kind of cancer) to the human population of a region, involves elaborate models of

- Environmental fate and transport.
- Exposure assessment.
- Toxicokinetics.
- Health effects.
- Population size, character and dynamics.
- Risk analysis.

For each one of these scientific disciplines, more than one model may be available (a toxicokinetics study, e.g., may be based on a single-compartment or a multi-compartment model). The various models above are inserted into the *TGIS*, which then evaluates their adequacy in terms of their ability to generate accurate space/time maps and lead to useful conclusions. ▪

In light of the above considerations, the *TGIS* specialists may find it appropriate under certain circumstances to adopt the view that

Scientific hypothesis testing is a model-dependent affair.

---

[6] One witnesses frequent examples of rather inadequate hypothesis testing nowadays. One day a group of experts announce to the media their confirmation of a hypothesis concerning the positive effects of a specific diet on a disease, only to be followed in a short period of time by the announcement from another group of experts claiming that their results show such a hypothesis to be totally invalid.

I.e., by successively incorporating into the *TGIS* the various models associated with the *same* hypothesis, the relative success of the corresponding conclusions (space/time maps, etc.) can provide evidence for or against the hypothesis.

Example 8.2. In order to test the hypothesis that, "Exposure to particulate matter is not a confounding factor for the temperature exposure-mortality association in the state of North Carolina during the winter of 1996", Christakos and Serre (2000) employed two different mathematical models. The first model did not account for the effect of the particulate matter distribution in space/time, whereas the second model did. As it turned out, the predictive mortality maps derived on the basis of the second model were no more accurate than those derived by the first model. This result provided some evidence that the hypothesis was correct. ▪

Often the model-dependent testing of a single hypothesis does not suffice but, instead, more than one considerably *different* hypothesis needs to be examined. In these cases

The TGIS-based scientific hypothesis testing can be viewed as a comparative affair

in the sense that the *TGIS* specialist seeks to demonstrate that a specific hypothesis is better supported by the scientific knowledge available compared to any of a series of competing hypotheses.

Example 8.3. The study of the geographotemporal evolution of an epidemic may involve a confrontation between different hypotheses. According to these hypotheses the main cause of the geographotemporal spread of the epidemic is due to

a The demographic growth of the host population.
b The global land-use changes.
c The collapse of geographical space.
d The worldwide effects of global warming.

Additional hypotheses may be based on joint interactions between the hypotheses (*a*) through (*d*) above. ▪

Comparative hypothesis testing often possesses a scientifically more sound basis than absolute hypothesis testing, which seeks to prove that a single hypothesis is confirmed or not by the evidence available. Indeed, in real world applications, any scientific hypothesis that passes all tests against the null hypothesis can still lose contests against competing hypotheses. As is shown in Fig. 8.1, the *TGIS* user often starts with a certain number of initial hypotheses (say, N_0), whereas some other hypotheses are generated during the process of reviewing the information provided by the *TGIS*. All the hypotheses are tested comparatively and several of them are excluded, thus leading to a decreasing number of hypotheses ($N_1 > \ldots > N_m > \ldots$) until the final hypothesis is confirmed. This final hypothesis could be (*a*) one of the initial hypotheses; (*b*) one of the updated hypotheses; or (*c*) some modification or combination of the above. During the comparative hypothesis testing, the *TGIS* user infers and tests several testable predictions generated by each one of the hypotheses across space and time.

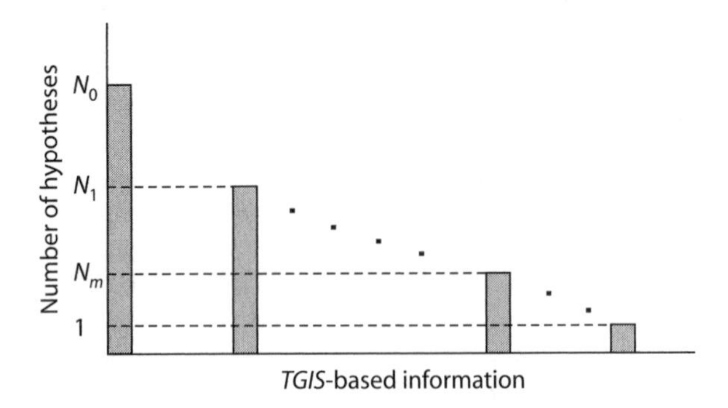

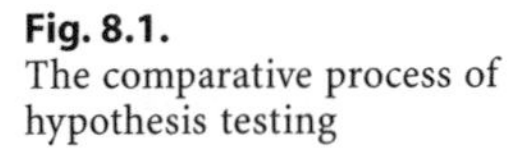

Fig. 8.1.
The comparative process of hypothesis testing

There exists a large variety of mathematical models in life support sciences that could potentially be used by *TGIS* specialists for scientific hypothesis testing purposes. A glance through a professional-level scientific journal is instructive on this point. The following major categories of models may be considered:

i. Analytical models.
ii. Computational (or numerical) models.
iii. Combinations of (*i*) and (*ii*).

In general, the larger and more complicated the numerical model is, the better fit it gives to the data but, also, the larger the number of parameters that need to be estimated. As a consequence, the choice of the appropriate model is not always a straightforward task (i.e., one does not necessarily choose the most sophisticated numerical model). When, e.g., insufficient site-specific data exist to calculate reliably the numerous parameters of the sophisticated model, it is doubtful that the analysis will generate more accurate maps than those produced by a simpler model involving a smaller but well-estimated number of parameters. Numerical solutions involve certain approximations and can require a considerable computational effort, but usually they can be obtained in complicated *TGIS* situations in which analytical solutions are not readily available. On the other hand, when possible to obtain, the analytical solutions have certain important features including the following: they provide the *TGIS* specialist with more insight into the behavior of the actual system than the numerical solution, and he can precisely calculate the outcome (solution) for any values of the input parameters. In many situations, a simpler analytical model can provide more insight and a deeper understanding of the underlying physical mechanisms than a series of numbers generated by numerical models. In the case of ecological modelling, e.g., Hilborn and Mangel (1997) maintained that

> Although the output of most models is numerical, the most influential models are the ones in which the numerical output is not needed to guide the qualitative understanding.

As far as hypothesis testing and explanation are concerned, we would like to suggest that this statement holds true in several other scientific disciplines, as well. In the next section we turn our attention to scientific explanation.

8.3 Scientific Explanation

Often, the goal of a hypothesis is to support scientific explanation. As a matter of fact, in the view of many researchers, only when a subject has reached the stage of explanation it is recognized as a science (Dunbar 1996). What exactly constitutes a scientific explanation is a topic that is open to various interpretations (deductive-nomological, inductive-statistical, etc.; see e.g., Salmon 1998). For the *TGIS* purposes of this volume, it suffices to view scientific explanation as a *process*. In particular,

Scientific explanation is a process in which a specific phenomenon is the effect of a set of initial/boundary conditions according to certain laws (physical principles, logical rules, etc.).

Often, scientific explanation is closely linked with *causation*. One then talks about causal explanation, the main ingredient of which is the search for causes. Thinking causally about a situation can improve the clarity of one's hypotheses and generate additional insights into the explanation process. Epistemically, it makes sense to distinguish between *deterministic* causation and *stochastic* causation[7]. Deterministic causal explanation might be possible only if all the relevant laws linking cause and effect were known in detail. Although very desirable, for a number of reasons (insufficient knowledge, complex mechanisms at work, lack of empirical support, etc.) such an approach is rarely materialized in practice[8]. Instead, many causality studies focus on the stochastic view which is concerned about some kind of *association* between cause and effect, rather than with a deterministic causal mechanism. There exist several stochastic causal explanation approaches that can benefit from *TGIS* modelling. Below we review some of these approaches.

A well-known explanation approach is based on the *inductive-statistical* concept according to which causality is manifested in terms of an almost perfect correlation between cause and effect (e.g., Dupre 1993). The inductive-statistical concept, however, suffers from a number of serious complications such as, while correlation is a symmetrical relation (e.g., if cigarette smoking is correlated with cancer, then cancer is correlated with smoking), causation is basically non-symmetrical (e.g., the fact that cigarette smoking causes cancer does not imply that cancer causes smoking).

Another approach is the so-called *probabilistic conditioning*, which assumes that the presence of a genuine cause must make the occurrence of its effect more likely than if it had not been present (Salmon 1998). Probabilistic conditioning is formalized in terms of conditional probabilities (e.g., the probability of the effect given the occurrence of the cause is greater than the probability of the effect given that the cause did not occur). Probabilistic conditioning is not free of substantial complications (see e.g., Skyrms 2000). One such complication is due to the so-called underlying common factor influence: It is possible that the assumed cause and the observed effect are both

[7] See discussion in Christakos (2001).

[8] For instance, while it had been the center of intensive research for decades, only recently has a mechanism been identified which explains how smoking can cause lung cancer (Denissenko et al. 1996).

consequences of some underlying factor (e.g., smoking and lung cancer may be both effects of an underlying genetically determined biological condition, in which case smoking is a symptom of having the bad gene rather than the cause of cancer). Thus, a rigorous application of the probabilistic conditioning approach should account for important factors, such as the different causally relevant contexts in which an exposure may occur and the different paths that may link an exposure to an effect.

A different view is based on the *critical predictability* concept discussed in Christakos and Hristopulos (1998) according to which, if the predictability of the effect is improved by taking into account *knowledge* about the cause, then the effect must be somehow associated with the specific cause. Given our current state of understanding of fundamental physical mechanisms, we may not be always able to explain this association rigorously in scientific terms, but the association nevertheless exists and cannot be ignored. Indeed, if we can obtain a better prediction of an important effect by using available knowledge about a potential cause, it would be irrational not to, and by doing so we acknowledge the importance of the specific cause in predicting the effect across space and time. In this sense, the problem with the possible existence of an underlying common factor is not as serious as in the case of the probabilistic conditioning approach above. In other words,

It is one thing to talk about abstract probabilities that do not necessarily have real counterparts and quite another thing to deal with real predictions in space and time that can be rigorously tested by means of the available observations.

Example 8.4. An application of the critical predictability concept in human exposure studies is the so-called *physico-epidemiologic prediction* (*PEP*) approach (Christakos and Serre 2000). *PEP* is based on the idea that the cognitive meaning of an cause-effect association lies in its potential for improved prediction. It combines logical and physical laws in a stochastic domain and assumes that the existence of a cause (exposure)-health effect association should lead to more accurate health effect predictions when information about both exposure and health effect is integrated than when only health data are used. By blending the physical measurements of exposure with the epidemiological features of effect in a space/time continuum, *PEP* accounts for various sources of uncertain knowledge, as well as for inter- and intra-subject variations for specified populations. ▪

As a matter of fact, depending on the case of interest, a *hierarchy* of scientific explanation levels can be associated with a particular phenomenon which, in turn, leads to different mathematical models of the situation. Explanations may involve models representing physical laws, scientific theories, or other forms of generalization. Actual reality can only be observed and explained in terms of these models on different levels. As is shown in Fig. 8.2, in most *TGIS* these modelling levels are not precise but rather useful pictures of the real world arranged vertically in accordance with their degree of generality (the sequence of explanation levels is denoted as "$\dots, N, N+1, N+2, \dots$"). Thus, a level with a higher degree of generality lies above one with a lower degree, in which case the former may be generated by improving the latter. A scientific law, e.g., belongs to a level which lies above the level of measuring instruments (allowing the observation of regions that are not accessible to the naked

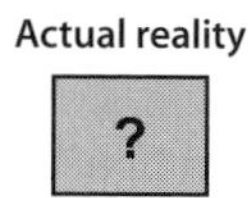
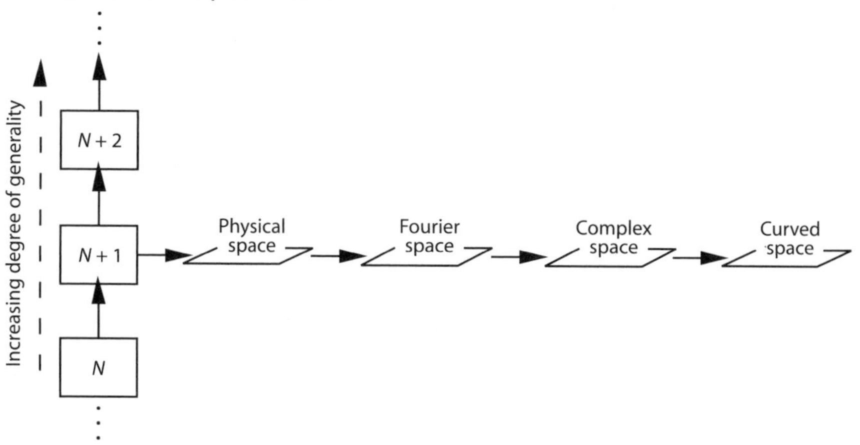

Fig. 8.2. Actual reality and *TGIS*-based scientific explanation levels

eye), which, in turn, lies above the level of everyday life observations (i.e., theory-free, direct sensory observations). A *TGIS* may involve a series of scientific theories with varying degrees of generality, each one of them explaining the phenomenon of interest at a different level. Thus, each one of the theories corresponds to a different mathematical model, and using the one or the other should depend on the goal of the *TGIS* investigation. Clearly, no upper limit exists for the number of explanation levels. In many cases, in addition to the vertical direction, a classification of models along the horizontal direction is also possible. Two or more models are equivalent when they belong to the same level vertically, but are developed in different spaces horizontally. Thus, a scientific law associated with a specific explanation level may be established in various representation frames, such as the physical space, a Fourier space, a complex space, or a curved space (Fig. 8.2). While they are structurally different from each other, all these frames are equivalent (from an information viewpoint) and belong to the same vertical position in the hierarchy of explanation levels. Depending on the explanation level considered, there are various ways to evaluate a scientific explanation. One can assess, e.g., whether a specific explanation (at a lower level in the hierarchy of Fig. 8.2) exemplifies a well-established general theory (of a higher level)[9]. Another powerful approach is to test if the site-specific predictions of the explanation produced by the *TGIS* are in agreement with an existing data base.

[9] According to a certain school of thought, unification is an important aim of scientific explanation (Rosenberg 2000). So, what makes an explanation of a specific phenomenon scientific is that the phenomenon is shown to be a special case of a more general process. The demand for unification puts logical deduction at the center of scientific explanation.

8.4 Geographotemporal Decision Making

Generally speaking, decision making is the process of generating decisions, usually under conditions of uncertainty. This process implies an allocation of resources in space and time, and its goal is to obtain as good a decision outcome as possible (e.g., Kleindorfer et al. 1993). To achieve such a goal, a number of elements must be combined:

- The necessary amount and quality of information.
- Powerful modelling tools.
- A decision process that is logically sound and physically meaningful.
- The efficient allocation of the resources available.

In many real-world applications, decision making depends on the *geographotemporal distribution* of the events under consideration[10]. In these circumstances, the *TGIS* has a substantial contribution in the development of an adequate decision making program (at a local, regional, national, or global level). Indeed,

i. Information gathering in space/time[11], and
ii. Mathematical tools (analytical and computational) for information processing,

are two of the *TGIS* constituents which provide powerful means to improve the input to the decision making process and, as a consequence, its final outcome. What type of mathematical tools the experts use will of course depend on the specific context of the investigation, and the questions that need to be answered. Clearly, many of the hypothesis testing and scientific explanation issues raised in the previous two sections are closely related to the decision making process. For instance, the use of models considered in hypothesis testing can offer important insights into various aspects of decision making. These kinds of insights could not have been obtained by means of descriptive approaches aiming to portray observed patterns of geographotemporal fields (natural, epidemiological, etc.)[12]. Birkin et al. (1996), e.g., demonstrated how modelling can improve public health planning and administration through the linkage of the geographical variations in health-care demand and the provision of health-care services in specific locations. Furthermore, the *TGIS*-based integration of hydrogeological, engineering, and economic models is required for water resources management purposes (McDonald and Key 1987).

Generally, an adequate decision making process exhibits the following three characteristics (see also Fig. 8.3):

[10] In the view of many experts the ultimate goal of *TGIS* is to offer support for making spatiotemporal decisions (e.g., Malczewski 1999).

[11] As a matter of fact, the main body of information used in decision making almost always comes from a large variety of sources (e.g., field data, laboratory measurements, remote sensing, computer simulations), which makes the *KB* synthesis capability of *BME* an especially suitable tool for the situation.

[12] Note that, descriptive approaches are mainly the focus of the classification and correlation stages of the methodology discussed in Sect. 8.1 above, whereas hypothesis testing and scientific modelling are considered in the cause-effect stage.

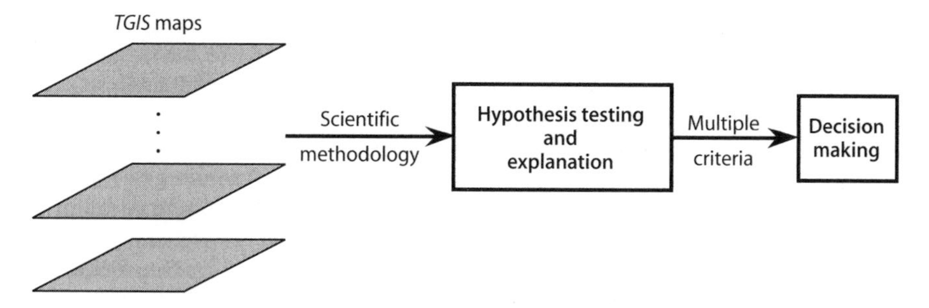

Fig. 8.3. *TGIS*-based geographotemporal decision making

- It considers several *TGIS-based alternatives* defined geographotemporally.
- It depends on the outcomes of the *hypothesis testing* and *explanation analysis*[13].
- It involves *multi-objective criteria*, which may themselves vary across space and time.

There are various settings of geographotemporal decision making within which *TGIS* tools could be used effectively. These settings may depend on a number of factors:

- The *issue* under consideration (subsurface contamination, spread of an epidemic, energy production and distribution, natural-resource management, ecological damage, etc.).
- The *natural environment* (space/time domain within which the relevant events take place, physical and human scales, time and space restrictions, etc.).
- The *economic conditions* (budgetary constraints, financial implications, trade-off, etc.).
- The *main players* (organizations, state and federal government, public interest groups, individuals, etc.).

The first example below demonstrates that the *TGIS* can play an important role in deciding which is the most appropriate strategy toward controlling or stopping the spread of an epidemic in geographical space and time. The second example deals with the determination of population risks due to environmental exposures.

Example 8.5. Cliff and Haggett (1989) and Haggett (2000) have presented different geographical control strategies aiming at preventing epidemic spread. These strategies include

a Local elimination (i.e., breaking – in a specific geographical area – the disease chain by vaccination).
b Defensive isolation (i.e., building a spatial barrier around a disease-free region).
c Offensive containment (i.e., the disease spread is halted and progressively eliminated by a combination of isolation and vaccination.
d Global eradication, which essentially combines the previous strategies (i.e., infected areas are progressively reduced in size, and the coalescence of disease-free regions eventually leads to the elimination of the disease). ▪

[13] As we shall see below, in some applications this part is called an *Intelligent Infobase*.

Example 8.6. Using site-specific chemical concentrations and standardized exposure variables, Hamilton and Viscusi (1999) calculated population risks for each one of the Superfund sites[14]. Then, they implemented a *TGIS* technology to combine these risks with maps of contaminant distributions, data on site boundaries, and the 1990 census in order to estimate the associated risks, in terms of expected numbers of cancers due to site contamination during a 30-year period. The total cost per cancer case averted at the site level was also estimated. On the basis of their "health risk-total cost" analysis, recommendations were made for use in effective decision making. ▪

Accurate space/time maps can help the decision maker differentiate local variations from widespread trends. This kind of information can be invaluable in the decision maker's efforts to identify potential causes of an event (e.g., water pollution). It can also be used effectively in risk evaluation and damage assessment due to potentially hazardous situations (natural disasters, etc.).

Example 8.7. In a recent study, Mynett (1999) suggested the use of *TGIS*-based flood simulation maps to assess potential damages to coastal areas, river-valleys, mountain and lowland regions, etc. These maps are combined with information on land use and investment, also available within the *TGIS*, to yield risk maps which are invaluable in decision making regarding the necessary protection measures to be taken before the event or, if the disaster cannot be avoided, the design of efficient evacuation plans of the region's population and livestock. ▪

The following example describes a project aiming at implementing an Egyptian air quality information and decision making system for the greater Cairo area, using *BME*-based advanced *TGIS* functions.

Example 8.8. Over the past few years, Cairo experienced a significant degradation in air quality due to rapid increases in population and industrialization. The goal of a study by Serre et al. (2000) was to establish the framework for an air quality *TGIS*, which would assist Egyptian decision-makers in implementing efficient air pollution abatement measures. The *TGIS* would provide interactive space/time images to monitor the air quality distribution in the greater Cairo area, to assess trends in air pollution and the efficacy of pollution abatement measures, and to take prevention measures when necessary. The databases used by the *TGIS* were provided by the Cairo Air Improvement Project (*CAIP*), which established a network of 34 stations to monitor ambient air levels of particulate matter (*PM*) and lead. Furthermore, the Environmental Information and Monitoring Program (*EIMP*) established a national monitoring network that includes 14 air quality monitoring stations in the greater Cairo area. These monitoring programs have recently become operational and the information collected provides the first air quality databases available on a continuous basis. In particular, one full year of *CAIP* data had been collected as of September 30, 1999. This constitutes the *CAIP* baseline year database, and serves as "benchmark" against which future monitoring will be compared to assess air quality trends. *EIMP* recently implemented a pro-

[14]The hazardous waste cleanup effort known as Superfund was launched by the U.S. government in 1980 in response to wide-spread public concern with hazardous wastes (see e.g., Chemical Week 1980).

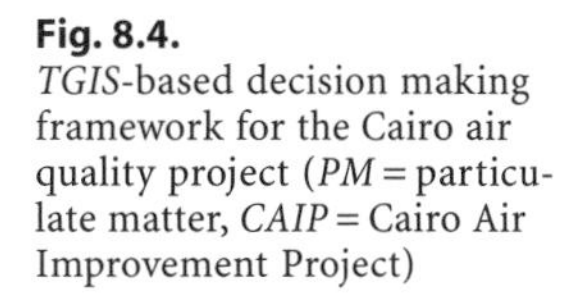

Fig. 8.4.
TGIS-based decision making framework for the Cairo air quality project (*PM* = particulate matter, *CAIP* = Cairo Air Improvement Project)

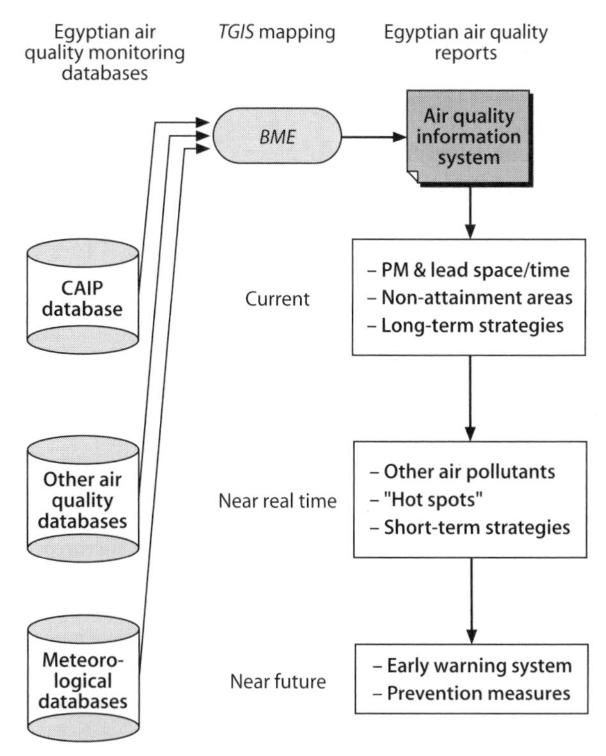

cedure to routinely provide its dataset to officials of Egyptian Environmental Affairs Agency (*EEAA*). As more data become available from these sources, a computerized *BME* system provides the framework for analyzing and modelling the air quality dataset. The *BMElib*-based decision making plan has three main levels (Fig. 8.4):

a *Current* stage: The Egyptian air quality databases of raw data collected by different institutions (with varying quality levels) serve as input to *BME*, which through the Egyptian Air Quality Information System generates on-demand pollutant space/time maps and reports of the Cairo air quality trends for decision-makers at *EEAA*.

b *Near real time* stage: Besides *CAIP*, the *EEAA* air quality databases provide information about other pollutants. These databases are naturally integrated into *BME* by means of soft information and, when these databases are collected electronically, *BME* will lead to a near-real time air pollution representation.

c *Near future* stage: Meteorological data are expected to become available at *EEAA*. By expanding the database to include additional information (cross-covariances between weather and air pollution parameters across space and time, etc.), the *BME* will lead to improved air quality assessment, thus opening the door to the implementation of an early warning system. ▪

In health sciences, *Disease Surveillance System Architecture* (*DSSA*) is an area that has undergone very rapid development in the last decade and continues to do so at an

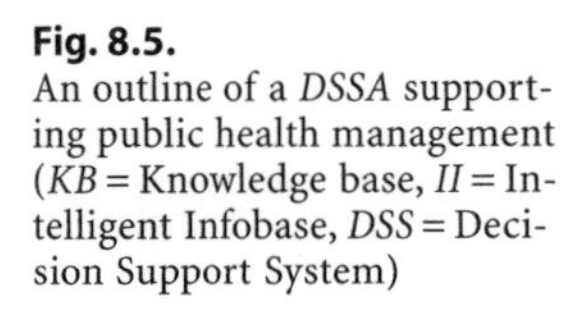

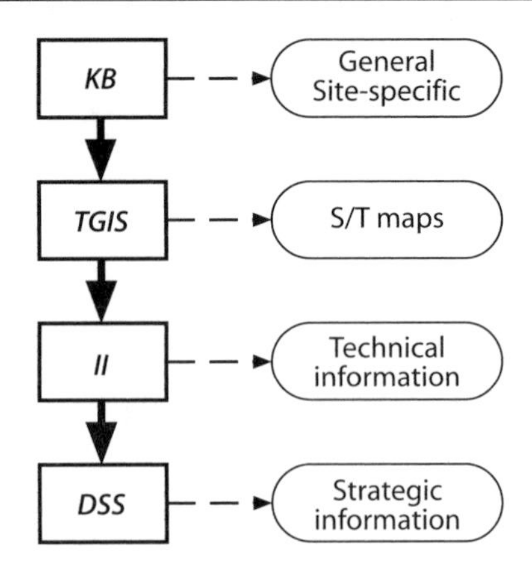

Fig. 8.5.
An outline of a *DSSA* supporting public health management (*KB* = Knowledge base, *II* = Intelligent Infobase, *DSS* = Decision Support System)

accelerating pace (see e.g., Forgionne et al. 2001). Most *DSSA* are built around *TGIS*, which, thus, play a very substantial role in health management-related activities at various levels (local, national, etc.; Tan 2001). As most studies indicate, an operationally efficient and decision-making effective DSSA should involve a number of crucial components, as follows (see also Fig. 8.5):

- The *Knowledge Base* (*KB*). As was discussed in previous chapters, a *KB* includes general physical theories, biological laws, empirical relationships, site-specific hard data, soft information, etc., all relevant to the disease under consideration. Clearly, the quality of DSSA heavily depends on the scientific quality of the *KB* used as input.
- The *TGIS*. The *TGIS* is at the heart of every *DSSA*. Enhanced with the advanced *BME*-based functions, a *DSSA* can generate accurate and informative space/time maps of the life support fields of interest (environmental exposures, gene frequency distributions, population dynamics, disease spread, etc.). The outcome maps (descriptive and predictive) of the *TGIS* are the main inputs to the subsequent *DSSA* steps (see also Fig. 8.3 above).
- The *Intelligent Infobase* (*II*). The *II* includes an organized repository of space/time maps (provided by the *TGIS*), models (conceptual, pictorial, and symbolic), quantitative indicators (incidence, mortality, etc.), and mathematical techniques. The *II* allows the *TGIS* users to describe, test, and explain spatiotemporal patterns, cause-effect associations, scientific hypotheses, and actions-reactions.
- The *Decision Support System* (*DSS*). This is a user-controlled framework that helps decision makers and policy designers to use the *II* above properly. One of the major goals of *DSS* is to turn the technical information provided by *II* into strategic information (performance indexes, economic evaluations, etc.). Based on this strategic information, the analytical reports generated by the *DSS* can greatly improve the effectiveness of top management.

An adequately designed *DSSA* provides a functional, integrated, and versatile architecture that is capable of supporting public health management at regional and national levels. Several successful applications can be found in the literature (see e.g., the Cancer Surveillance System Architecture discussed in Forgionne et al. 2001).

8.5 Prelude

Generally speaking, an adequate *TGIS* should possess a sound epistemic structure guaranteeing internal consistency (free of logical contradictions), offer the proper means to incorporate various knowledge sources, make possible the generation of accurate and informative predictions, and play a constructive role in scientific hypothesis testing, explanation and decision making. As a result of the multi-facet demands and the changing conditions of real world applications, a realistic methodology may combine elements from more than one philosophical schools.

In the view of many experts, the rapidly evolving *TGIS* technology follows a path the main features of which can be represented by a Kuhnian schema of scientific methodology (see Sect. 8.1 above). Based on the general description of such a schema outlined in Klee (1997), Fig. 8.6 presents a flowchart of the corresponding two major methodological frames of a *TGIS*: The normal and the revisionistic frames. The *normal TGIS* frame offers a reasonable representation of an established paradigm, which involves a considerable amount of technical and often highly detailed work at the classification, correlation and cause-effect levels outlined in Sect. 8.1 above. Within the normal frame, *TGIS* practitioners are engaged in inquiry under the direction of the paradigm. No established paradigm, however, is perfect. Eventually, due to the intrinsic incompleteness of paradigms there will be a few research results that do not fit the established paradigm (these results are usually called "anomalies"). As a consequence, paradigm shifts would occur at later stages which establish new boundaries and allow the solution of certain fundamental problems that were previously unsolvable (see *revisionistic TGIS* frame in Fig. 8.6). To be sure, the revisionistic process takes time to

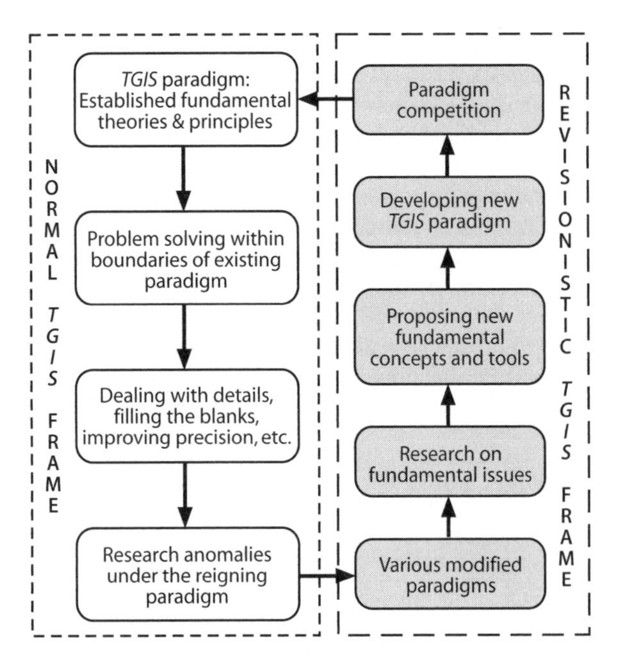

Fig. 8.6. A flowchart representation of *TGIS* methodological frames

be completed, because no expert who has built his career on the basis of the old paradigm (within which he was originally trained) is willing to see his work questioned under the new paradigm (which may render his work irrelevant, inadequate or even worthless).

At this point, it may be worth noticing that, an influential group of historians of science often adds to the above flowchart a substantial social factor which emphasizes the difference between the way science ought to be practiced vs. the way it really turns out to be practiced by many practitioners. In the words of Klee (1997):

> The historical record showed that scientists were often vain, conceited, willing to fudge data, willing to argue unfairly, willing to use larger power sources in society to further their own interests and beliefs.

In addition, the brutal way the "Old Guard" treats younger practitioners who dare to propose a new paradigm has nothing to do with the pursuit of scientific truth. Klee describes the situation as follows:

> The Old Guard stoops to trying to thwart the careers of the younger revolutionaries. The Old Guard tends to control access to research money, to publication space in the professional journals of the field, and to academic appointments at institutions of higher learning. It can get nasty, petty, and downright unpleasant.

Sample (1996) argues further that in reality the research design often suits the desires and agendas of the researcher rather than the subject(s) of the research. In some cases, the guardians of the reigning paradigm form a kind of a "cult" enlisting mindless allegiance and acquiescence. The cult members are devoted to attacking any new ideas which, in their view, threaten the vested interests of the cult. It is, perhaps, true that a considerable percentage of practicing scientists closely resemble the above description, and the sociology-based critics have every right to point out the "cult-like" methods of this group of scientists[15]. However, often the same critics seem to fall into a paradoxical trap: while they start by pointing out the inappropriate practices of the above group of scientists, they then proceed to legitimize these practices by proposing ideological frameworks (cultural, social and political) that appear to fit closely the group's activities they criticized in the first place. For example, on the one hand they stigmatize the intervention of a researcher's social agenda into the design and excecution of his research project and on the other hand they propose an ideological framework that allows for such an agenda, promotes fashionable "political correctness" over scientific value and substance, and gives in to the demands of nonrational social factors. In our view, an ideological framework should instead support the practices of the honest and rational scientists who are dedicated to the pursuit of knowledge and truth, regardless of irrelevant outside pressures.

Returning to more pleasant affairs, there is no doubt that *TGIS* offers a considerable amount of valuable information to the decision maker, in terms of space/time

[15]Sometimes these methods are encouraged, even institutionalized by Federal Government agencies. A recent article in *Washington Post* titled "EPA urged to improve its scientific research" (Thursday, June 15, 2000) is enlightening, in this respect. The article refers to a report issued by the National Research Council regarding the US EPA's (Environmental Protection Agency) highly questionable policy of scientific research funding. According to the author of the article, "... the report is also critical of EPA's peer-review process, noting that current policy allows the same individual who manages a project to serve as the peer-review leader for that project." *O tempora, o mores*!

correlation graphs, maps and other kinds of visualization, technical reports, etc. Having access to information is not necessarily the same as having the acumen it takes to interpret it. By definition, decision making is an *interdisciplinary* approach, which requires input from several scientific disciplines.

To be effective, decision making should be based on critical thinking and a deeper understanding of the life support processes (physical, epidemiological, ecological etc.) at work, in order to be able to distinguish between content and form, sense and sensibility, and weighty thought and ponderous words.

Furthermore, one must not forget that the *TGIS* tools used in decision making are instruments for knowledge and, thus, of power. This can have crucial social, political, financial, etc. implications. Public health and environmental decision making, e.g., can be a heavily political process, in the best as well as in the worst senses of the term (Dale and English 1999). Nevertheless, whatever the potential implications might be one needs to decide. Indeed, a state of indecision is itself a decision not to act. And the consequences of non-decision can be as serious (and in some cases, even more serious) than those of making a specific decision.

References

The real voyage of discovery consists not in seeking
new landscapes, but in having new eyes.
M. Proust

Aerts D (1985) A possible explanation for the probabilities of quantum mechanics and example of a macroscopical system that violates Bell inequalities. In: Mittelstaedt P, Stachow E-W (eds) Recent developments in quantum logic. Wissenschaftsverlag Bibl. Inst., Zurich, pp 235–249

Anselin L (1995) SpaceStat tutorial (with SpaceStat version 1.80 user's guide). Regional Research Instit., West Virginia Univ., Morgantown

Arlinghaus SL (ed) (1995) Practical handbook of spatial statistics. CRC Press, New York

Armstrong DM (1983) What is a law of nature? Cambridge Univ. Press, Cambridge

Asher HB (1983) Causal modeling. Sage Publ., Newbury Park, London

Baiamonte A (1996) An equity model for locating environmental hazardous facilities in densely populated urban areas. Dept. of Geography, Hunter College, City Univ. of New York

Bailey RG (1995) Ecosystem geography. Springer, Berlin Heidelberg New York

Bakhavlov N, Pansenko F (1989) Homogenization: Averaging processes in periodic media. Kluwer Academic, Dordrecht

Bennett R (1979) Spatial time series. Pion, Nondon

Birkin M, Clarke G, Clarke M, Wilson A (1996) Intelligent GIS. Geoinformation International, Cambridge

Bloschl G, Sivapalan M (1995) Scale issues in hydrological modelling: A review. In: Kalma JD, Sivipalan M (eds) Scale issues in hydrological modelling. J. Wiley & Sons, New York, pp 9–48

Bogaert P (2001) Comparison between BME for categorical data and indicator (co) Kriging. Stochastic Environmental Research & Risk Assessment, submitted

Bogaert P, Christakos G (1997) Spatiotemporal analysis and processing of thermometric data over Belgium. J Geophysical Research-Atmospheres 102(D22):25831–25846

Bogaert P, Serre M, Christakos G (1999) Efficient computational BME analysis of non-Gaussian data in terms of transformation functions. IAMG99, Proceed. of 5th Annual Confer. of the Internat. Assoc. for Mathematical Geology, August 6–11, 1999, Trondheim, Norway

Bolstad PV, Gessler P, Lillesand TM (1990) Positional uncertainty in manual digitized map data. Intern J GIS 4:399–412

Bras RL, Rodriguez-Iturbe I (1985) Random functions in hydrology. Addison-Wesley, Reading

Bretherton F, Davis R, Fandry C (1976) A technique for objective analysis and design of oceanographic experiments applied to MODE-73. Deep Sea Research 23:559–592

Burrough PA (1986) Principles of geographical information systems for land resources assessment. Clarendon Press, Oxford

Burrough PA, Frank AU (eds) (1996) Geographic objects with indeterminate boundaries. Taylor & Francis, London

Burrough PA, McDonnell RA (1998) Principles of geographical information systems. Oxford Univ. Press, Oxford

Bury KV (1975) Statistical models in applied science. J. Wiley & Sons, New York

Cassettari S (1993) Introduction to integrated geo-information management. Chapman & Hall, London

Chemical Week (1980) Superfund: how it will work, what it will cost? Chemical Week Dec. 17:38

Cherkassky Y, Mulier F (1998) Learning from data: concepts, theory and methods. J. Wiley & Sons, New York

Chiles J-P, Delfiner P (1999) Geostatistics: modeling spatial uncertainty. J. Wiley & Sons, New York

Choi K-M, Serre ML, Christakos G (2001) Spatiotemporal BME analysis and mapping of mortality distribution in California. In: Proc. of 53rd Session of the Inter. Statistical Inst., Seoul, S. Korea, 22–29 August, 2001

Chrisman NR (1997) Exploring geographical information systems. J. Wiley & Sons, New York

Christakos G (1980) Soil structure interaction: A laboratory apparatus for studying the behavior of soils under dynamic loading. Research Rep. Foundation Engineering, University of Birmingham

Christakos G (1985) Recursive parameter estimation with applications in earth sciences. J Mathematical Geology 17(5):489–515
Christakos G (1990) A Bayesian/maximum-entropy view to the spatial estimation problem. J Mathematical Geology 22(7):763–776
Christakos G (1991) Some applications of the Bayesian maximum entropy concept in geostatistics. Fundamental Theories of Physics, Invited paper, Kluver Acad. Publ., pp 215–229
Christakos G (1992) Random field models in Earth sciences. Academic Press, San Diego
Christakos G (1998a) Spatiotemporal Information Systems in soil and environmental sciences. Geoderma 85(2–3):141–179
Christakos G (1998b) While God is raining brains, are we holding umbrellas? The role of modern geostatistics in spatiotemporal analysis and mapping. Keynote lecture. In: Buccianti A, Nardi G, Potenza R (eds) IAMG98, Proceed. of 4th Annual Confer. of the Internat. Assoc. for Mathematical Geology v. 1, De Frede Editore, Naples, Italy, pp 33–53
Christakos G (2000) Modern spatiotemporal geostatistics. Oxford Univ. Press, New York, 2nd edition (2001)
Christakos G (2001) Stochastic modelling in human exposure. In: El-Shaarawi AH, Piegorsch WW (eds) Encyclopedia of Environmentrics. J. Wiley & Sons, New York
Christakos G, Bogaert P (1996) Spatiotemporal analysis of springwater ion processes derived from measurements at the Dyle Basin in Belgium. IEEE Trans Geosciences and Remote Sensing 34(3):626–642
Christakos G, Hristopulos DT (1996) Stochastic indicators for waste site characterization. Water Resoures Research 32(8):2563–2578
Christakos G, Hristopulos DT (1998) Spatiotemporal environmental health modelling: A tractatus stochasticus. Kluwer Acad. Publ., Boston
Christakos G, Kolovos A (1999) A study of the spatiotemporal health impacts of ozone exposure. J Exposure Analysis & Environ Epidemiology 9(4):322–335
Christakos G, Li X (1998) Bayesian maximum entropy analysis and mapping: A farewell to kriging estimators? Mathematical Geology 30(4):435–462
Christakos G, Papanicolaou V (2000) Norm-dependent covariance permissibility of weakly homogeneous spatial random fields. Stochastic Environmental Research & Risk Assessment 14(6):1–8
Christakos G, Serre ML (2000a) Spatiotemporal analysis of environmental exposure-health effect associations. J Exposure Analysis & Environ Epidemiology 10(2):168–187
Christakos G, Serre ML (2000b) BME analysis of spatiotemporal particulate matter distributions in North Carolina. Atmospheric Environment 34:3393–3406
Christakos G, Vyas V (1998) A composite spatiotemporal study of ozone distribution over eastern United States. Atmospheric Environment 32(16):2845–2857
Christakos G, Hristopulos DT, Bogaert P (2000) On the physical geometry hypotheses at the basis of spatiotemporal analysis of hydrologic geostatistics. Adv Water Resour 23:799–810
Christakos G, Serre ML, Kovitz J (2001) BME representations of particulate matter in the State of California. J Geophysical Research-D 106(D9):9717–9732
Chung KL (1968) A course in probability theory. Academic Press, San Diego
Clarke KC (1986) Advances in geographic information systems. Computers, Environment and Urban Systems 10(3/4):175–184
Cliff AD, Hagget P (1988) Atlas of disease distributions: Analytic approaches to epidemiologic data. Basil Blackwell, London
D'Or D, Bogaert P, Christakos G (2001) Applications of BME to soil texture mapping. Stochastic Environmental Research & Risk Assessment 15(1):87–100
Dale VH, English MR (1999) Tools to aid environmental decision making. Springer, Berlin Heidelberg New York
Daley R (1991) Atmospheric data analysis. Cambridge Univ. Press, Cambridge
Dangermond J (1984) A classification of software components commonly used in GIS. In: IGU Commission on Geographical Data Sensing and Processing (ed) Proceed. of the U.S./Australia Workshop on Design and Implementation of Computer-Based GIS, Amherst, MA, pp 70–91
Davis B (1996) GIS: A visual approach. Onword Press, Santa Fe
Denissenko MF, Pao A, Tang M, Pfeifer GP (1996) Preferential formation of benzo[a]pyrene adducts at lung cancer mutational hotspots in p53. Science 274:430–432
Deutsch CV, Journel AG (1992) Geostatistical software library and user's guide. Oxford Univ. Press, Oxford
Diggle PJ, Tawn JA, Moyeed RA (1998) Model-based geostatistics. Appl Statist 47:299–350
Dunbar R (1996) The trouble with science. Faber and Faber Ltd., London
Dupre J (1993) The disorder of things. Harvard Univ. Press, Cambridge
Entekhabi D, Rodriguez-Iturbe I (1994) Analytical framework for the characterization of the space-time variability of soil moisture. Advan in Water Resour 17:35–45

ESRI-Environmental Systems Research Institute (1990) Understanding GIS: The ARC/INFO method. Redlands, CA
Fisher P (ed) (1995) Innovations in GIS-2. Taylor & Francis, London
Forgionne GA, Gangopadhyay A, Adya M, Tan JKH (2001) Data warehousing, data mining, and integrated health decision support systems: A comprehensive cancer surveillance system architecture. In: Tan JKH (ed) Health management information systems. Aspen Publ., Inc., Gaithersburg
Frieden BR (1991) Probability, statistical optics, and data testing. Springer, Berlin Heidelberg New York
Gandin LS (1963) Objective analysis of meteorological fields. Gidrometeorolog. Izdat., Leningrad, USSR (English translation, Israel Program of Scient. Transl., Jerusalem, Israel, 1965)
Ghil M, Cohn S, Tavantzis J, Bube K, Isaacson E (1981) Applications of estimation theory to numerical weather prediction. In: Bengtsson L, Ghil M, Kallen E (eds) Dynamic meteorology: Data assimilation methods. Springer, Berlin Heidelberg New York
Glymour C (1981) Theory and evidence. Princeton University Press, Princeton
Goodchild MF (1989) Modeling error in objects and fields. In: Goodchild MF, Gopal S (eds) Accuracy of spatial data bases. Taylor & Francis, London, pp 107–113
Goovaerts P (1997) Geostatistics for natural resources evaluation. Oxford Univ. Press, New York
Guye E (1922) L'Evolution Physico-Chimique. Chriton, Paris
Haggett P (2000) The geographical structure of epidemics. Clarendon Press, Oxford
Haggett P, Cliff AD, Frey A (1977) Locational models. Halstead Press, New York
Hamilton JT, Viscusi WK (1999) Calculating risks. M.I.T. Press, Cambridge
Heuvelink GBM (1998) Error propagation in environmental modelling with GIS. Taylor & Francis Lts., London
Hilborn R, Mangel M (1997) The ecological detective: Confronting models with data. Princeton Univ. Press, Princeton
Holmes EE (1997) Basic epidemiological concepts in a spatial context. In: Tilman D, Kareiva P (eds) Spatial ecology. Princeton Univ. Press, Princeton
Houghton J (1997) Global warming. Cambridge Univ. Press, Cambridge
Houlding SW (2000) Practical geostatistics. Springer, Berlin Heidelberg New York
Hristopulos DT, Christakos G (1999) Renormalization group analysis of permeability upscaling. Stochastic Environmental Research & Risk Assessment 13(2):131–160
Hristopulos DT, Christakos G (2001) Calculations of non-Gaussian multivariate moments for BME analysis. Mathematical Geology 33(5):543–568
Jenks GF, Caspall FC (1971) Error on choropleth maps: Definition, measurement, reduction. Annals of the Assoc of American Geographers 61:217–244
Johnson DH, Dudgeon DE (1993) Array signal processing. Prentice Hall, Englewood Cliffs
Jones R (1965) An experiment in non-linear prediction. J Appl Meteor 4:701–705
Journel AG (1983) Non-parametric estimation of spatial distributions. Math Geol 15(3):445–468
Journel AG (1989) Foundamentals of geostatistics in five lessons. Amer. Geophys. Union, Washington
Kailath T (1981) Lectures on Wiener and Kalman filtering. Springer-Verlag, New York
Kanevski M, Arutyunyan R, Bolshov L, Demyanov V, Chernov S, Savelieva E, Timonin V, Maignan M, Maignan MF (1999) Mapping of radioactively contaminated territories with geostatistics and artificial neural networks. In: Linkov I, Schell WR (eds) Contaminated forests. Kluwer Acad. Publ., Dordrecht, The Netherlands, pp 249–256
Khrennikov A (1997) On the physical interpretation of negative probabilities in Prugovecki's empirical theory of measurement. Canadian J Physics 75(5):291–298
Kitanidis PK (1997) Introduction to geostatistics. Cambridge Univ. Press, Cambridge
Kitchin R, Tate NJ (2000) Conducting research into human geography. Prentice Hall, Harlow
Klee R (1997) Introduction to the philosophy of science. Oxford Univ. Press, New York
Kleindorfer PR, Kunreuther HC, Schoemaker PGH (1993) Decision sciences: An integrative perspective. Cambridge University Press, Cambridge
Klir GJ, Yuan B (1995) Fuzzy sets and fuzzy logic: Theory and applications. Prentice Hall, Upper Saddle River
Kolmogorov AN (1939) Sur l'interpolation et extrapolation des suites stationnaires. Comptes Rendus Académie des Sciences, Paris 208:2043–2045
Kolovos A, Christakos G, Serre ML (2000) Incorporation of physical laws and other forms of knowledge in spatiotemporal prediction of hydrologic processes. EOS Trans. Amer. Geoph. Union, San Francisco, CA, Dec. 15–19, 2000
Kraak MJ, Ormeling FJ (1996) Cartography: Visualization of spatial data. Longman, Essex
Krumhardt BA, Wirth DM (1999) Experiences in environmental science. Bellwether-Cross Publ., East Dubuque
Kuhn TS (1962) The structure of scientific revolutions. University of Chicago Press, Chicago
Lakatos I (1976) Proofs and refutations. (Worrall J, Zahar EG (ed)) Cambridge Univ. Press, Cambridge

Langran G (1992) Time in geographic information systems. Taylor & Francis, London
Laurini R, Thompson D (1995) Fundamentals of spatial information systems. Academic Press, San Diego
Lemons DS (1997) Perfect form-variational principles, methods, and applications in elementary physics. Princeton Univ. Press, Princeton
Limb PR, Meggs GJ (1995) Data mining – tools and techniques. British Telecom Technology Journal 12(4):32–41
Linhart H, Zucchini W (1986) Model selection. J. Wiley & Sons, New York
MacEachren AM, Taylor DRF (eds) (1994) Visualization in modern cartography. Pergamon/Elsevier Sci. Ltd, Oxford
Maitin IJ, Klaber KZ (1993) Geographical information systems as a tool for integrating air dispersion modelling. In: GIS/LIS Proceed., Amer. Soc. of Photogrammetry and Remote Sensing, Nov. 2–4, Minneapolis, pp 466–474
Malczewski J (1999) GIS and multicriteria decision analysis. J. Wiley & Sons, New York
Matern B (1960) Spatial variation. Medd Fran Stat Skogsf 49(5)
Matheron G (1965) Les Variables Régionalisées et Leur Estimation. Masson, Paris
Matheron G (1973) The intrinsic random functions and their applications. Adv Appl Prob 5:439–468
MathWorks Inc. (1998) MatLab, the language of technical computing, using MATLAB version 5. The Mathwork Inc. http://www.mathworks.com, Natick
Maurer BA (1994) Geographical population analysis: Tools for the analysis of biodiversity. Blackwell Sci., Oxford
Mausner JS, Kramer S (1985) Epidemiology. W.B. Saunders Co., Philadelphia
McDonald AT, Kay D (1986) Water resources: Issues and strategies. Longman, London
Michael J (2000) Anxious intellects. Duke Univ. Press, Durham
Miller KS (1980) An introduction to vector stochastic processes. R.E. Krieger Publ. Co., Huntington
Muller JC (1987) The concept of error in cartography. Cartographica 24:1–15
Mynett AE (1999) Hydroinformatics and its applications at Delft hydraulics. J Hydroinformatics 1(2): 83–102
Newton RG (1997) The truth of science. Harvard University Press. Cambridge
Nikias CL, Petropulu AP (1993) Higher-order spectra analysis. Prentice-Hall, Upper Saddle River
Olea RA (1999) Geostatistics for engineers and Earth scientists. Kluwer Acad. Publ., Boston
Omatu S, Seinfeld JH (1981) Filtering and smoothing for linear discrete-time distributed parameter systems based on Wiener-Hof theory with application to estimation of air pollution. IEEE Trans on Systems, Man, and Cybernetics 11(12):785–801
Oreskes N (2000) Why predict? Historical perspectives on prediction in earth science. In: Sarewitz D, Pielke RA Jr., Byerly R Jr. (eds) Prediction: Science, decision making and the future of nature. Island Press, Washington D.C., pp 23–40
Osleeb JP, Kahn S (1999) Integration of geographic information. In: Dale VH, English MR (eds)Tools to aid environmental decision making. Springer, Berlin Heidelberg New York
Parrish D, Cohn S (1985) A Kalman filter for a two dimensional shallow-water model: Formulation and preliminary experiments. Office Note 304, National Meteorological Center, US Dept. of Commerce, National Weather Service, Washington D.C.
Petersen D (1973) A comparison of the performance of quasi-optimal and conventional objective objective analysis schemes. J Appl Meteor 12:1093–1101
Petersen DP, Middleton D (1965) Linear interpolation, extrapolation, and prediction of random space-time fields with a limited domain of measurement. IEEE Trans. on Information Theory 11(1):18–30
Pfaff SP, Peteet DM (2001) Generating probabilities in support of societal decision-making. Eos Trans Amer Geophysical Union 82(20):222, 225
Popper KR (1934) Logik der Forschung. Springer, Vienna
Popper KR (1962) The logic of scientific discovery. Hutchinson, London
Rhind DW, Green NPA (1988) Design of a geographic information system for a heterogeneous scientific community. Intern J GIS 2:171–189
Rivoirard J (1994) Introduction to disjunctive Kriging and non-linear geostatistics. Clarendon Press, Oxford
Robinson AH, Morrison JL, Muehrcke PC, Kimerling AJ, Guptill SC (1995) Elements of cartography, 6th ed. J. Wiley & Sons, New York
Robinson VB (1988) Some implications of fuzzy set theory applied to geographical databases. Computers & Geosciences 16(7):857–872
Rosenberg A (2000) Philosophy of science. Routledge, New York
Salmon WC (1998) Causality and explanation. Oxford Univ. Press, New York
Sample PL (1996) Beginnings: participatory action research and adults with developmental disabilities. Disability and Society 11(3):317–332

Sarewitz D, Pielke RA Jr., Byerly R Jr. (2000) Prediction: Science, decision making and the future of nature. Island Press, Washington D.C.
Sauer CO (1952) Agricultural origins and dispersals. American Geographical Society, New York
Serre ML (2001) Numerical aspects of the implementation of the BME method using soft information. CASEnews 1(2), Center for the Advanced Study of the Environment, University of North Carolina, Chapel Hill
Serre ML, Bogaert P, Christakos G (1998) Computational investigations of Bayesian maximum entropy spatiotemporal mapping. In: Buccianti A, Nardi G, Potenza R (eds) IAMG98, Proceed. of 4th Annual Confer. of the Internat. Assoc. for Mathematical Geology 1:117–122, De Frede Editore, Naples, Italy
Serre M, Christakos G, Howes J, Gamal A (2000) Powering an Egyptian air quality information system with the BME space/time analysis toolbox: Results from the Cairo baseline year study. Proc. of GeoEnv2000 (3rd European Confer. on Geostatistics for Environmental Applications), Avignon, France, Nov. 22–24, 2000
Sheppard E, Leitner H, McMaster RB, Tian H (1999) GIS-based measures of environmental equity: Exploring their sensitivity and significance. J Exposure Analysis & Environmental Epidemiology 9:18–28
Sinton D (1978) The inherent structure of information as a constraint to analysis: mapped thematic data as a case study. In: Dutton G (ed) Harvard Papers on GIS, vol. 7. Addison-Wesley, Reading
Sivia DS (1996) Data analysis. Clarendon Press, Oxford
Skyrms B (2000) Choice & chance, 4th ed. Wadsworth, Stamford
Spitz K, Moreno J (1996) A practical guide to groundwater and solute transport modeling. J. Wiley & Sons, New York
Steede-Terry K (2000) Integrating GIS and the global positioning system. Environmental Systems Research Inst., Inc., Redlands
Steinberg EK, Kareiva P (1997) Challenges and opportunities for empirical evaluation of "spatial theory". In: Tilman D, Kareiva P (eds) Spatial ecology. Princeton Univ. Press, Princeton, pp 318–332
Tan JKH (ed) (2001) Health management information systems. Aspen Publ., Inc., Gaithersburg
Tanizaki H (1993) Nonlinear filters. Springer, Berlin Heidelberg New York
Thiebaux HJ, Pedder MA (1987) Spatial objective analysis. Acad. Press, San Diego
Tulku T (1990) Knowledge of time and space. Dharma Publ., Oakland
Voltz M, Lagacherie P, Louchart X (1997) Predicting soil properties over a region using sample information from a mapped reference area. European J Soil Science 48:19–30
Wang C (1993) Sense and nonsense of statistical inference. Marcel Dekker, Inc., New York
Weisstein EW (1999) CRC coincise encyclopedia of mathematics. Chapman & Hall/CRC, Boca Raton
Wiener N (1949) Time series. M.I.T. Press, Cambridge
Wittgenstein L (1961) Notebooks 1914–16. Wright GH von, Anscombe GEM (eds), Blackwell, Oxford
Wold H (1938) A study in the analysis of stationary time series. Almqvist & Wiksell, Stockholm
Wolf PR, Brinker RC (1989) Elementary surveying, 8th ed. Harper & Row Publ., New York
Wood J (1996) Scale-based characterisation of digital elevation models. In: Parker D (ed) Innovations in GIS. Taylor & Francis Ltd., London, pp 163–175
Wurman RS (2001) Information anxiety 2. QUE, Indianapolis

Index

Nullumst iam dictum quod non dictum sit prius.
Terence

T

U

V

W

Springer

Druck: Strauss Offsetdruck, Mörlenbach
Verarbeitung: Schäffer, Grünstadt